T0313851

Optoelectronic Materials and Technology in the Information Age

Related titles published by The American Ceramic Society

Dielectric Materials and Devices
Edited by K.M. Nair, Amar S. Bhalla, Tapan K. Gupta, Shin-Ichi Hirano,
Basavaraj V. Hiremath, Jau-Ho Jean, and Robert Pohanka
©2002, ISBN 1-57498-118-8

Functionally Graded Materials 2000 (Ceramic Transactions Volume 114)
Edited by K. Trumble, K. Bowman, I. Reimanis, and S. Sampath
©2002, ISBN 1-57498-110-2

The Magic of Ceramics
By David W. Richerson
©2000, ISBN 1-57498-050-5

Electronic Ceramic Materials and Devices (Ceramic Transactions, Volume 106)
Edited by K.M. Nair and A.S. Bhalla
©2000, ISBN 1-57498-098-X

Ceramic Innovations in the 20th Century
Edited by John B. Wachtman Jr.
©1999, ISBN 1-57498-093-9

Dielectric Ceramic Materials (Ceramic Transactions, Volume 100)
Edited by K.M. Nair and A.S. Bhalla
©1997, ISBN 0-57498-066-1

Advances in Dielectric Ceramic Materials (Ceramic Transactions Volume 88)
Edited by K.M. Nair and A.S. Bhalla
©1996, ISBN 1-57498-033-5

Hybrid Microelectronic Materials (Ceramic Transactions Volume 68)
Edited by K.M. Nair and V.N. Shukla
©1995, ISBN 1-57498-013-0

Ferroic Materials: Design, Preparation, and Characteristics (Ceramic Transactions, Volume 43)
Edited by A.S. Bhalla, K.M. Nair, I.K. Lloyd, H. Yanagida, and D.A. Payne
©1994, ISBN 0-944904-77-7

Grain Boundaries and Interfacial Phenomena in Electronic Ceramics
(Ceramic Transactions Volume 41)
Edited by Lionel M. Levinson and Shin-ichi Hirano
©1994, ISBN 0-944904-73-4

For information on ordering titles published by The American Ceramic Society, or to request
a publications catalog, please contact our Customer Service Department at 614-794-5890
(phone), 614-794-5892 (fax), <customersrvc@acers.org> (e-mail), or write to Customer
Service Department, 735 Ceramic Place, Westerville, OH 43081, USA.

Visit our on-line book catalog at <www.ceramics.org>.

Volume 126

Optoelectronic Materials and Technology in the Information Age

Proceedings of the Optoelectronic Materials and Technology in the Information Age symposium at the 103rd Annual Meeting of The American Ceramic Society, held April 22–25, 2001 in Indianapolis, Indiana.

Edited by

Ruyan Guo
The Pennsylvania State University

Allan Bruce
Lucent Technologies

Venkatraman Gopalan
The Pennsylvania State University

Basavaraj Hiremath
Tyco Submarine Systems Ltd.

Burtrand Lee
Clemson University

Man Yan
Lucent Technologies

Published by
The American Ceramic Society
735 Ceramic Place
Westerville, Ohio 43081
www.ceramics.org

Proceedings of the Optoelectronic Materials and Technology in the Information Age symposium at the 103rd Annual Meeting of The American Ceramic Society, held April 22–25, 2001 in Indianapolis, Indiana.

COVER PHOTO: SEM micrograph showing the formation of prismatic Zn_2SiO_4 particles is courtesy of Chulsoo Yoon and Shinhoo Kang, and appears as figure 3D in their paper "Effects of Starting Compositions on the Phase Equilibrium in Hydrothermal Synthesis of Zn_2SiO_4:Mn^{2+}," which begins on page 59.

ISBN 978-1-57498-134-6

For information on ordering titles published by The American Ceramic Society, or to request a publications catalog, please call 614-794-5890, or visit <www.ceramics.org>.

Contents

Electro-Optic and Ferroic Materials in Optoelectronic Applications

Preface

The explosive growth of the information industries has been enabled by the availability and performance of key optoelectronic components including optical fibers, lasers, amplifiers, filters, modulators, detectors, display, and storage devices. These are based on transparent, electro-optic, nonlinear-optical, luminescent, and emissive glass and ceramic materials. The continuing growth and evolution in this area presents exciting opportunities for the investigation and development of new optoelectronic materials and devices.

This symposium was co-sponsored by the Electronics Division, the Glass and Optical Materials Division, and the Basic Science Division of The American Ceramic Society (ACerS). Organizers represented the three divisions in an effort to provide a unique interdisciplinary forum within a strong materials community for technical exchange on optoelectronic materials, device applications, and system development. The symposium was held April 22–25, 2001, in Indianapolis, Indiana, during the 103rd Annual Meeting of The American Ceramic Society. Sessions were focused on the latest achievements on both materials and device technologies, which can lead to further advances in the communication, data storage, display, biomedical, and defense industries. Forty-seven presentations including ten invited talks, twenty-three contributed talks, and fourteen posters were presented in the two and one-half day symposium program.

The content of this Ceramic Transactions volume comprises the proceedings of the symposium. The papers featured in this volume are divided into three parts: inorganic phosphor, display, and solid-state lighting materials; novel synthesis of amorphous and semiconducting optoelectronics; and electro-optic and ferroic materials in optoelectronic applications.

The organizers of this symposium and editors of this volume acknowledge and appreciate the contributions of the speakers, symposium session chairs, authors, manuscript reviewers, and ACerS program coordinators for making this symposium a successful one. Organizational and financial support from the three co-sponsoring divisions also is acknowledged.

Ruyan Guo
Allan Bruce
Venkatraman Gopalan
Basavaraj Hiremath
Burtrand Lee
Man Yan

Inorganic Phosphor, Display, and Solid State Lighting Materials

LUMINESCENCE OF LONG-TIME ORDERED GaP:N

Sergei L. Pyshkin
Institute of Applied Physics, Academy of Sciences,
Academy Str.5, MD2028, Kishinev, Moldova

ABSTRACT
The results of investigations of long-time ordering of nitrogen atoms along P sites in 35 years ago grown GaP single crystals and related phenomena in N impurity-bound excitons are presented. It was shown that during this period impurity redistribution due to the respective substitution reaction with the energetic barrier 1.0-1.2 eV at room temperature and normal pressure leads to regular distribution of N along anion sites. A combine molecular beam and laser assisted method of growth of GaP thin epitaxial films periodically doped with N atoms is proposed. Possible properties of high density bound excitons in these epitaxial films have been discussed taking into account that the ordered N impurity-bound exciton phase of high density gives new opportunities for appearance of various non-linear optic phenomena, accumulation, conversion and transport of light energy.

INTRODUCTION
It is known that Gallium Phosphide (GaP) is highly prospective both for various applications (photoreceivers, light emitters, electroluminescent displays, power diodes etc.) as well as a model material for investigations of fundamental properties of semiconductors. In particular, GaP doped with nitrogen N or GaP:N is a unique object for generation, investigation and application of bound exciton system which is mostly interesting in the field of high density of excitons bound to ordered N-impurity superlattice.

Due to considerable efforts in single crystal growth during my graduate and postgraduate courses (1963-1966) at State Technical University, A.F. Ioffe Physico-Technical Institute (St.Petersburg, Russia) and Moldavian Ac. Scie. Institute of Applied Physics now, probably, only I am the owner of the unique collection of over 35 years "matured" single crystals of pure and doped Gallium Phosphide. I seems that very interesting long-time changes periodically for years observed in these crystals give a unique opportunity to develop new generation of electronic devices as well as physics of ordered excitonic state. The program of the long-time monitoring of these changes is planned for 50 years since 1965 through app. each 10 years and the last monitoring was fulfilled during 1991-1993 at Cagliari University, Italy, so, the next period will be initiated since 2001.

The main goals of the presented talk and paper are (i) to discuss the results of investigations of long-term ordering of nitrogen atoms along P sites in 35 years ago grown GaP single crystals and related phenomena in N impurity-bound excitons, (ii) to propose the methods of growth of GaP thin epitaxial films periodically doped with N atoms and (iii) to discuss possible properties of bound excitons in these epitaxial films.

RESULTS AND DISCUSSION

The single crystals were grown in 1963-1966 from super high pure Ga and P components at the temperatures excluding contamination of the crystals in the growth process using a special furnace with very high accuracy automatic control of temperature that leads to ultimately possible high quality of crystals grown in laboratory conditions.

The crystals were doped by various chemical elements that to ultimately purify them (Zr), to introduce donors D (VI group: S, Se, Te), acceptors A (II group: Zn), centers of irradiative D-A recombination (Zn-Te, Zn-O), deep traps (elements of Fe group: Fe, Ni, Co), electron or hole traps with a giant cross-section - centers of bound exciton creation (N and Bi), centers of radiative recombination screened from temperature and crystal field influences (rare-earth elements Sm, Gd or La) as well as the impurities increasing the quantum output of irradiation up to 100% (Sm + N).

The single crystals grown at very steep cooling ($5°C$ per hour) from weak (5 at.%) melted solution of P in Ga with a small quantity of GaN (N impurity-bound excitons) or Zr (free excitons) have the most efficient luminescence of bound excitons and perfect quality [1].

Parameters and properties of the noted above GaP crystals as well as of their ternary analogue $CdIn_2S_4$ - temperature dependencies of conductivity and Hall effect, photoconductivity and luminescence at various temperatures and levels of optical excitation were investigated in the years of their preparation as well as during the years after. The mostly considerable results obtained for these years are the development of growth technology for perfect and free of contamination doped crystals, investigation of influence of crystallization conditions on morphology and quality of crystals [1], discovering of impact ionization of bound excitons [2], stimulated emission of radiation through indirect optical transition [3], "giant" shift of luminescent bands in dependence on intensity of exciting light and increase of quantum output of irradiation [4, 5], multi-quantum optical transitions with participation of 2 - 7 photons or 2 photons and a phonon participation in an elementary light absorption act [6, 7], collective properties in high density bound exciton system [8] as well as long-time ordering (LTO phenomenon) of impurities and host elements in these crystals for 35 years aged at normal conditions [9-12].

N impurity-bound excitons in GaP:N are under investigation around the world since early 60th [13, 14]. Note, that N isoelectronic trap has a giant capture cross-section; so only full extraction of nitrogen traces from the crystals at chemical interaction between N and Zr can result to creation of free excitons. Luminescent spectra of excitons and excitonic molecules bound to isolated N atoms ($r_{NN} > 5$ nm, where r_{NN} - the average spacings between N impurities), as well as excitons bound to NN pairs ($r_{NN} = 1.2$ nm and less) have been investigated to the middle of 70th. To the end of 70-th main properties of bound excitons of high density $\{(n_1 \times a_1^3)^{1/3} \cong 1$, where n_1 and a_1 - bound exciton concentration and Bohr radius respectively, $a_1 = 5$ nm, $n_1 = N_0 \cong 10^{18}$ cm^{-3}\} have been also studied experimentally [8, 15] and theoretically [16].

The noted above investigations show that all groups of luminescent particles such as free excitons, excitons and excitonic molecules bound to single N atoms, excitons bound to NN pairs have their own specific luminescent spectra consisting of zero phonon line and its phonon replica. One can select experimental conditions in which only one group of particles gives the main contribution in luminescence. However, due to a stochastic distribution of impurities inside the crystals each of the groups presents and can be experimentally distinguished.

For the first time it was shown that the system of high density bound excitons can be considered as a solid excitonic phase like to an inverted alkali metal and consisting of a net of negatively charged heavy nuclei (N atom + captured electron) which interacts with a free hole gas [8]. Since electrons are being captured by a short-range potential of impurity states while holes are being bound by a long-range Coulomb interaction, N isoelectronic impurities localize electrons much stronger than holes. Thus, creation of this phase were recognized by a specific changes of luminescence spectrum at high density of bound excitons (Fig. 1) as well as by p-type photoconductivity of highly optically excited GaP:N in photo-Hall measurements compared to always n-type dark conductivity of these crystals (Fig.2) [8, 9, 12].

It was also demonstrated that in spite of random distribution of N impurities along P sites just after preparation of GaP:N crystals, disposition of impurities only in anion sites as well as intense Auger recombination between the bound excitons disposed at a too short distance to each other ($<a_1$) give some initial ordering to the solid excitonic phase.

In 1980 M.Combescot and C.B. a la Guillaume [16] have made a brilliant theoretical generalization of the experimentally observed collective properties published for the first time in 1974 by S.L.Pyshkin and L.Zv. Zifudin [8]. The

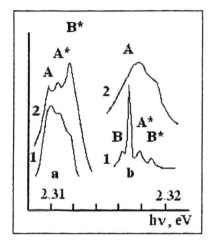

Fig. 1. Evolution of GaP:N luminescence spectra at $(n_1 \times a_1^3)^{1/3}$ less than (1) or equal to 1 (2). a. 4.2K. b. 80K. A,B - excitons; A*,B* - biexcitons.

Fig. 2. Kinetics of photoconductivity (solid) and photo-Hall effect (dotted) in GaP:N. 80K. $(n_1 \times a_1^3)^{1/3}$ less than 1 (1, 2) or equal to 1 (3,4).

authors [16] have proved that there exist two successive phase separation for an electron-hole system in a semiconductor having isoelectronic impurities within a certain density range. One is between bound excitons and a hole plasma with electrons pinned on the impurities, and the other one is between this hole plasma and the usual electron-hole plasma.

Thus, 25 years ago a solid excitonic phase with some inclinations of ordering like to an excitonic crystal has been predicted and discovered. In contrast to the high density system of free excitons admitting only liquid phase and do not existing in solid state, the solid bound excitonic phase has heavy nuclei with small zero vibrations like an usual crystal with metallic bonds, rather high critical temperature destroying this new phase (21 meV) and the lifetime 10^{-7} s [17] which is enough that to observe its creation.

A new stage of experiments with the aged (grown in the middle of 60th) GaP:N crystals has started at the beginning of 90-th when a dramatic evolution of their luminescent and Raman spectra (Figs. 3 and 4) has been revealed.

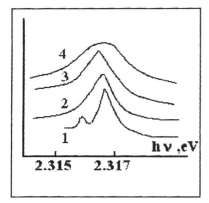 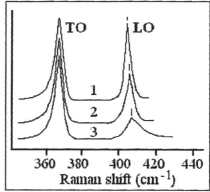

Fig. 3. Evolution of the luminescent non-phonon line of bound exciton A as a function of N concentration. 15K. 1-3: ordered crystals. 4: unordered. 1-4: $N_o = 10^{17}$; 10^{18}; 10^{19} and 10^{18} N in cubic cm respectively.	Fig.4. Evolution of LO lattice phonon energy with N concentration in ordered GaP:N according to Raman scattering at 300K. Excitation by Ar^+-laser at 514 nm. Concentration of N grows in 3 orders from 1 to 3 (app. 10^{19} N in cubic cm).

1. Very intense in fresh crystals doped by 10^{19} cm^{-3} and more of N (average r_{NN} = 3 -5 nm) NN_1 - NN_{10} luminescence of excitons bound to NN pairs with the spacing r_{NN} from 0.345 nm (NN_1) to 1.219 nm (NN_{10}) is completely absent in the same crystals 25 years after their growth. This experimental result means that the degree of impurity ordering in the aged crystals is rather high compared to the fresh ones - now divergences from the average r_{NN} do not exceed 10% in the crystals doped by 4 x10^{19} cm^{-3} against 100% and more in the fresh crystals. Instead of the NN_1 -NN_{10} lines, the luminescence spectra of ordered crystals develop a broad uniform band the maximum of which shifts to the low energy region as the nitrogen concentration increases.

2. Zero phonon line A and its phonon replica of single N impurity-bound excitons in the aged crystals shift their positions with the concentration of N impurities according to the relation between of A position and N_0 concentration:

$$E_{NN} = E_N - \beta r_{NN}^{-3} \ (1),$$

where E_N is the A line position at $r_{NN} \rightarrow \infty$, E_{NN} is the same at some non-zero nitrogen concentration and $\beta = 13$ if E_N, E_{NN} are measured in eV and r_{NN} in Å [10]. Meanwhile, compared with the ordered Ga:N the fresh crystals demonstrate only broadening of the lines when the nitrogen concentration increases that also confirms ordered disposition of impurities with equal spacing r_{NN} in the aged crystals (Fig.3). An increase of N concentration in ordered GaP:N system leads also to increase, for instance, of LO lattice phonon energy because nitrogen replaces 2 times heavier phosphorus in ordered N superlattice (Fig.4).

3. All the details of luminescent spectra of N impurity-bound single excitons can be observed in the aged crystals more clearly than in fresh ones. It concerns to a very low background of the exciton lines as well as to their small halfwidth and distinct position in the spectrum. Thus, the observed long-term ordering is actual also for host constituents of crystal lattice. In general, GaP:N crystals are considerably more perfect after 25 years than just after their growth. Note, that nearly the same results were obtained by us for $CdIn_2S_4$, which during this period turned their partly inverse spinel structure T_d^2 into the normal spinel O_h^7 [10, 11].

Thus, it was shown that the impurity redistribution since the middle of 60-th up to the beginning of 90-th due to the respective substitution reaction with the energetic barrier 1.0-1.2 eV at room temperature and normal pressure gives regular disposition of N along anion sites.

A solid bound exciton phase is absolutely new phenomenon. Previous results [8, 9, 12] only give arguments for its existence without a proper investigation of its properties such as crystal structure, luminescence, conductivity and charge transport, non-linear effects etc. Meanwhile, it is clear that these investigations will be very interesting and useful for various applications. Particularly, if in atoms the characteristic parameter for a non-linearity appearance is the first ionization potential of the order of 10eV, the same critical parameter for an exciton crystal, the energy of exciton, is 3 order less or 10^{-2} eV only that gives an unique opportunity to generate various non-linear optical effects at very low exciting light intensity. This excitonic phase of high density gives also new opportunities for accumulation, conversion and transport of light energy. Now we plan to apply for growth and monitoring of ordered GaP:N films new growth technique, such as combined molecular beam and laser vacuum epitaxies (MBE and LVE) with various modern *in situ* and *ex situ* diagnostics of growing film [18-21] and use of in-built ion lithography. The properties of these films will be investigated in comparison with bulk ordered GaP:N crystals.

New stage in solving of the problem of crystal state of bound excitons depends on the opportunities, which give modern methods of thin film preparation and their doping by impurities. Really, with the recent progress in film growth there is no necessity to wait an ordering for decades. For instance, superlattice from GaP/GaP:N with the period of the order of the Bohr radius, which is equal to 5 nm, can be prepared by MBE or in combination with the Laser Vacuum Epitaxy [18]. Preparation of two-dimensional net of N impurities along pure GaP film is rather difficult, but it is also possible with the help of ion lithography or atomic force microscope (AFM) built into a growth chamber. Of course, nowadays this technique is a frontier of our technological possibilities, but during the nearest 2-3 years very important results will be obtained also in this direction.

CONCLUSION

For years, properties of bound excitons were not so attractive and understandable as properties of free excitons. It seems, however, excitons (single, pairs, excitonic molecules) bound to N impurity superlattice after their proper investigations will be important object for solid-state physics and its applications in the nearest future. Especially it concerns to collective properties of high density bound excitons, excitonic molecules as well as phase excitonic transitions in ordered GaP:N system. Except initial theoretical generalization and a set of experimental data in this field to the recent moment there is not any complete picture of the observed phenomena.

Thus, the next goals of this protracted for many years study are (i) molecular beam and/or laser assisted growth in ultra-high vacuum chamber equipped with in situ diagnostics of advanced heterostructures for EL displays as well as thin epitaxial GaP films periodically doped by N atoms, (ii) investigation of the heterostructures for EL displays; (iii) investigation of long-term ordering in 35 years ago grown GaP single crystals and related phenomena in N-impurity bound exciton system in the films with ordered disposition of N impurities.

Particularly, it will be necessary to determine by optical or other methods identity period as well as the perfection both of impurity and excitonic superlattices in aged crystals and epitaxial films, to investigate also in comparison with fresh crystals peculiarities of luminescence, Raman scattering, photo-Hall effect, generation of second harmonic of light and other non-linear optical phenomena in presence of the induced excitonic superlattice.

Both the investigation and further application of the long-term ordering (LTO) are of great interest because besides the fundamental scientific reasons the phenomenon gives a unique opportunity to considerably improve perfection of artificially grown crystals as well as to develop new or increase known useful properties applicable for semiconductor device making, probably, for watch and jeweler industries.

In particular, the ordered high density bound exciton system with app. Bohr radius distance between the excitons is a crystalline excitonic phase having unique highly useful properties. It is obvious also that wares fabricated from the aged crystals are considerably more tolerant to degradation than their fresh analogues. It was shown by us also that long-term ordering in some cases could be replaced by a short-time temperature treatment of fresh crystals.

Results of investigation of the discussed phenomena obtained with the unique collection of ordered GaP single crystals will give a new approach to fabrication of integral circuitries for opto- and microelectronics: it will concern choice of aged crystals having very stable parameters in time (low degradation of devices on their base) instead of just grown as well as realization of giant light capacity and low threshold non-linear effects in dense ordered bound excitonic phase. GaP films periodically doped by N atoms can be artificially grown by molecular beam and/or laser assisted epitaxies and will be used in new generation of mesascopic light emitters, transducers and receivers.

Note, that GaP:N crystals are only a small part of this collection of "matured" crystals and properties of GaP doped by other noted above impurities have not been investigated in details up to now, so new projects could be generated during this work.

It's a pity, but the most important for application final stage of this study of ordered GaP crystals coincides with the total crises all over the post-Soviet territory that has destroyed in the first place fundamental investigations and high technologies for opto- and microelectronics. So, I hope that addressing my appeal to collaboration I'll meet kind response and support of world scientific community and possible sponsors.

REFERENCES
1. N.A. Goryunova, S.L. Pyshkin, A.S. Borschevsky, S.I.Radautsan, G.A.Kaliujnaia, Yu.I. Maximov and O.G.Peskov, "Influence of Impurities and Conditions of Crystallization on Growth of Platelet GaP Crystals", (in Russian), *J. Rost crystallov*, **8(2)** 84-92 (1968).
2. B.M. Ashkinadze, I.P. Kretsu, S.L. Pyshkin, S.M.Ryvkin and I.D.Yaroshetsky, "Effect of Electric Field and Temperature on Exciton Radiation Intensity in GaP", *J. Soviet Physics-Solid State* **10(12)** 2921-2922 (1968).
3. S.L. Pyshkin, "Stimulated Emission in Gallium Phosphide", *J. Soviet Physics-Doklady* , **219(6)** 13451974

4. S.L. Pyshkin, I.A.Damaskin, S.I. Radautsan and V.E. Tazlavan, "An Anomalous Shift of Luminescent Bands in Some Semiconductors", *J. Sov Phys- JETP Letters,* **18(4)** 239-240, (1973)

5. S.L. Pyshkin, "On Luminescence of GaP:N:Sm Single Crystals", *J. Sov Phys - Semicond.,* **8** 1397-1399 (1974).

6. S.L. Pyshkin, N.A.Ferdman, S.I.Radautsan, V.A.Kovarsky, E.V. Vitiu, "Many-Quantum Absorption in Gallium Posphide", *J. Opto-electronics* **2** 245-249 (1970).

7. I.A. Damaskin, S.L. Pyshkin, S.I. Radautsan, V.E. Tazlavan, "Multi-quantum photoconductivity in $CdIn_2S_4$," 1973 *J. Opto-electronics* **5** 405-410 (1973).

8. S.L. Pyshkin and L.Zv. Zifudin, "Excitons in Highly Optically Excited Gallium Phosphide", *J. Lum.* **9** 302-310 (1974).

9. S.L. Pyshkin, A. Anedda, F. Congiu and A. Mura , "Luminescence of the GaP:N ordered system", *J. Pure Appl. Opt.* **2** 499-503 (1993).

10. V.A. Budyanu, I.A.Damaskin, Val.P. Zenchenko, S.L. Pyshkin, S.I. Radautsan, A.A.Suslov, and V.E.Tezlevan, "Processes of long-lasting ordering in crystals with a partly inversed spinel structure", *J. Sov. Phys. Dokl.* **35(4)** 301-303 (1990).

11. I.A. Damaskin, Val.P. Zenchenko, S.L. Pyshkin, S.I.Radautsan, I.M. Tiginyanu, V.E.Tezlevan, and V.N. Fulga, "Raman spectra of $CdIn_2S_4$ with different cation-sublattice ordering", *J. Sov. Phys. Dokl.* **35(12)** 1064-1065 (1990).

12. Sergei Pyshkin and Alberto Anedda A, "Time-dependent behaviour of antistructural defects and impurities in $CdIn_2S_4$ and GaP", *Proc. Int. Conf. Ternary and Multinary Comp.,* Salford 1997, *Inst. Phys. Conf. Ser., Section E,* No **152** 785-788 (1998).

13. D.G.Thomas, M.Gershenzon, and J.J.Hopfield, *J. Phys. Rev.,* **131,** 2397 (1963)

14. D.G.Thomas and J.J.Hopfield, "Isoelectronic Traps Due to Nitrogen in Gallium Phosphide", *J. Phys. Rev.,* **150,** 680-689 (1966).

15. R.Schwabe, F.Thusselt, H.Weinert, and R.Bindemann, *J. Lum.,* **18/19,** 537 (1979).

16. M.Combescot and C. Benoit a la Guillaume, *J. Phys. Rev. Lett.,* **44(3),** 182 (1980).

17. B.M.Ashkinadze, A.I.Bobrysheva, E.V. Vitiu, V.A.Kovarsky, A.V. Lelyakov, S.A. Moskalenko, S.L.Pyshkin, S.I.Radautsan, "Some NonlinearOptical Effects in GaP", *Proc. IX Int. Conf. Phys Semicond.* (1968, Moscow) vol. **1,** 189-193 (1968).

18. S.Pyshkin, S.Fedoseev, S.Lagomarsino, and C.Giannini, "Preparation and structural properties of some III-V semiconductor films grown on (100) oriented Si substrates", *J. Appl. Surf. Scie.,* **56-58,** 39-43 (1992).

19. Pyshkin S L, Heterostructures (CaSrBa)F$_2$ on InP for Optoelectronics, 1995, *Report to The United States Air Force European Office of Aerospace R&D on SPQ-94-4098.*

20. Pyshkin S L, Grekov V P, Lorenzo J P, Novikov S V, Pyshkin K S, "Reduced Temperature Growth and Characterization of InP/SrF$_2$/InP(100) Heterostructure", *Physics and Applications of Non-Crystalline Semiconductors in Optoelectronics, NATO ASI Series,* 3.High Technology, Vol.36, p. 468 (1996).

21. S.L. Pyshkin, CdF$_2$:Er/CaF$_2$/Si(111) Heterostructure for EL Displays, 1997, *Report to The United States Air Force European Office of Aerospace R&D on SPQ-97-4011.*

RADIOLUMINESCENT GLASS BATTERY

M.M. Sychov
St. Petersburg Institute of Technology
19 Moscow prospect
St. Petersburg, Russia 199001

K.E. Bower
TRACE Photonics Inc.
20 North 5th street
Charleston, IL 61920

A.G. Kavetsky
Khlopin Radium Institute
28 Second Moorinsky prospect
St. Petersburg, Russia 194021

V.M. Andreev
Ioffe Institute
26 Politechnicheskaya street
St. Petersburg, Russia 194021

ABSTRACT
Proof-of-principle experiments are described for improving radioisotope-based power supplies using scintillation glass. Radioisotope batteries have higher energy density than chemical batteries and a much longer life. In our experiments, low-light threshold photovoltaic cells produced electricity when exposed to Tb-activated scintillating glass waveguides sandwiched between tritium beta-sources. Advantages of indirect radioisotope energy conversion to electricity are that the semiconductor is not exposed to highly ionizing, and therefore damaging, radiation; glass is a relatively stable matrix for radioisotope immobilization; and the glass waveguide increases the semiconductor's conversion efficiency while reducing its size. Radiation source self-absorption is a limiting problem with both indirect and direct conversion approaches. The overall efficiency measured in this work was less than the estimated 1% potential for an optimized indirect conversion system.

INTRODUCTION
Radioluminescent light sources (RLS) are used in safety and military signage. One of the intriguing uses of RLSs is in nuclear batteries for electrical micropower generation. In such batteries, the energy of radioactive decay is first converted into light, and light is then converted into electricity with the use of sensitive photovoltaic cells. Because of the two steps of energy conversion, this indirect approach is contrasted with "direct conversion" where isotope radiation is

directly converted into the electricity across alpha- or betavoltaics. Nuclear batteries might be designed to utilize even mixed nuclear waste.

Typical radioluminescent light sources are internally phosphor-coated glass bulbs filled with the gas of the hydrogen isotope, tritium (^{3}H). Another type of RLS may be based on thin waveguides made of scintillating glass. Alternating layers of scintillating glass and tritium have been used in our work. Light is concentrated due to waveguiding effect and emitted from the glass edges. This paper describes fabrication and testing of an indirect conversion nuclear battery.

ADVANTAGES OF NUCLEAR BATTERIES:

Chemical batteries have a relatively low ratio of stored energy to volume and weight. For example, the zinc-mercury battery ratio is 0.55 W-h/cm^3 while the nuclear battery ratio can exceed 50 W-h/cm^3 [1]. Another shortcoming of the chemical battery is a short lifetime, while the nuclear battery can be used for decades, depending on the radioisotope. Chemical batteries are controlled by chemical reaction kinetics which are sensitive to temperature. Therefore, chemical batteries do not function well at temperature extremes.

Compared to chemical batteries, nuclear batteries have many significant advantages:

- Continuous Operation Life of Over Ten Years
- High Energy Density
- Slow Reduction of Power Output based on the Decay Rate
- Low Weight
- Small Size
- Wide Interval of Operating Temperatures with No Performance Loss at Very Low Temperatures
- High Reliability
- Might Use a Feedstock Now Considered Worthless

While limited in instantaneous power output, nuclear batteries may be useful in a variety of products including medical implants, sensors, remote applications, and other microelectronic devices [2]. Nuclear batteries may be used to charge capacitors to provide needed power surges in such applications such as remote transceivers. The best power match appears to be in the emerging field of micro-electromechanical systems (MEMS) with very low power requirements, miniature size, and long operation time. The total MEMS market is expected to exceed $1B in the year 2005 with exponential growth in the following decade [3]. The availability of a reliable power supply will enhance the development of this important emerging technology.

CHOICE OF ISOTOPE

To be suitable for use in nuclear batteries, radioisotopes should possess intermediate decay rates with the necessary useful life; high specific activity; a minimum of gamma radiation which has a high penetration depth and requires complex protection; low cost; availability, acceptable regulatory and safety restrictions, and a convenient fabrication technology. Properties of some suitable isotopes are summarized in Table 1 [4]. Alpha emitters provide the highest energy density, but degrade crystalline and amorphous materials [5,6]. Tritium is an acceptable isotope due to a moderate decay time, availability, and low cost. The soft betas permit its use without extraordinary radiation shielding. Therefore for our experiments, tritium was used. Tritium is usually used in a gaseous form as $^3H_2$. Gaseous tritium limits miniaturization, so bound tritium is preferable, even though it greatly increases the radiation source self-absorption. A useful solid form is titanium tritide, obtained by saturation of a fresh thin layer of titanium by tritium with annealing at 450-500 °C.

Table I. Isotopes suitable for nuclear battery application

Isotope	Decay type	Maximum energy of decay particles, MeV	Half life, years	Gamma emission
^{85}Kr	Beta	0.67	10.4	0.5%
^{204}Tl	Beta	0.763	3.8	No
^{147}Pm	Beta	0.224	2.65	No
^3H	Beta	0.0186	12.34	No
^{63}Ni	Beta	0.067	100	No
^{241}Am	Alpha	5.5	435	Weak
^{238}Pu	Alpha	5.6	86.4	Weak

BATTERY DESIGN

One of the major barriers to nuclear battery commercialization is self-absorption of the ionizing radiation in both direct and indirect conversion variants, requiring large semiconductor surface areas to collect the tiny radiation flux. Since the semiconductor cost dominates the battery cost, the large semiconductor surface area increases the cost to power. Direct conversion also has the central problem of semiconductor radiation damage. This suggests the value of indirect conversion designs in which the semiconductor is not exposed to highly ionizing radiation. Use of scintillating waveguides improves the indirect design by increasing the intensity of light incident on the photovoltaic cell. The nuclear battery efficiency increases since the photovoltaic efficiency increase with higher flux. This design significantly reduces photovoltaic surface area which is essential for miniaturization and cost reduction. Multilayer structures needed for practical energy output are cheaper to fabricate when one of the repeating components, in this case the photovoltaic cell, is removed from the structure and placed outside (see Figure 1). Such a design better protects the photovoltaic from radioisotope diffusion, which in the case of tritium, degrades the semiconductor both radiolytically and chemically.

The most efficient scintillators are ZnS-based phosphors and alkali halides like NaI(Tl). However, A_2B_6 compounds are not transparent enough to emitted light. Alkali halides possess good clarity but are hygroscopic. Fabrication of thin alkali halide waveguides by polishing is costly. Plastic scintillation is an established technology but the radiation stability of the organic matrix is too poor. Scintillating glass based waveguides are nearly ideal for this application. Glasses have scintillation efficiency of several percent, radiation stability, and good waveguide properties. Glass waveguides may be fabricated as fibers [7] as well as RF-magnetron sputtered to form flat waveguides suitable for multilayer glass/isotope stack design [8]. Figure 1 shows proposed design of radioluminescent glass based battery.

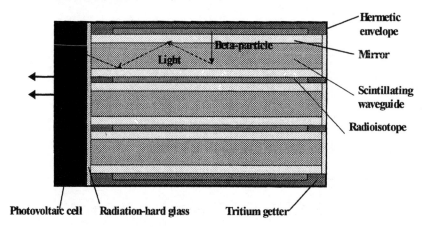

Figure 1. Cross-section of the concentrator battery design.

Hermetic sealing of the radioisotope with a tritium getter such as 1,4-bis(phenylethynyl)benzene (DEB) prevents accidental isotope leakage. Scintillation glass waveguides are coated with thin mirrors (or high difference refractive index barrier) and a radioisotope source. The shape of the waveguides may be round, square or flat. In the experimental setup, tritium as titanium tritide generates beta radiation, soft beta particles penetrate the scintillating glass to

Optoelectronic Materials and Technology

generate visible photons, which are guided to the emitting edge. A radiation-hard borosilicate glass window optically couples the waveguide and photovoltaic, while preventing radioisotope diffusion to the semiconductor. The spectral response of the photovoltaic cell should be matched to the emission maximum of the scintillating material. The radioisotope may be advantageously incorporated into the waveguide. For example, promethium oxide may be incorporated into the glass.

Summary of the advantages of the waveguide scintillation glass nuclear battery:

- The semiconductor is not exposed to highly ionizing radiation.
- Glass withstands high temperatures for deposition of Ti tritide.
- Conversion efficiency is increased by increase of light flux on the surface of the photovoltaic.
- Miniaturization and decrease of cost achieved by reduction of needed size of photovoltaic.
- Waveguides may be fabricated in different shapes and with thickness of absorption range of tritium betas allowing further miniaturization.
- Possibility to incorporate isotope into the scintillating material to improve power density and efficiency.

EXPERIMENTAL

Samples of 1-mm thick Tb-activated scintillation glass were obtained from Collimated Holes Inc. This glass has improved radioluminescence efficiency and radiation stability due to additional doping with Gd and Ce. To validate the stability of the glass with tritium exposure, a glass disk was sandwiched between two titanium tritide sources and the light output was measured for 31 days with the use of International Light 1700 photometer with SHD033 detector (silicon photodiode).

For better matching of the photovoltaic to the glass' emission, its photoluminescent properties were studied. The maximum on the excitation spectrum of the glass is 270 nm, so this wavelength was used for excitation during collection of the photoluminescence spectrum. There are secondary excitation peaks at 245 and 315 nm.

For this experiment with photovoltaic cells, polished glass was cut into four 21x5 mm pieces of a millimeter thickness and assembled into a stack. Each piece was sandwiched between two titanium tritide sources on a steel substrate having 100 um thickness. A highly sensitive (low-light threshold) AlGaAs photovoltaic cell with a 4x4 mm aperture was utilized. Its response curve matched the emission maximum of the glass and the cell had a very low leakage current of about 10^{-12} A/cm^2 at U = 10 mV.

RESULTS AND DISCUSSION

As seen from Figure 2, light output from the scintillation glass pinned between tritiated titanium was stable within the accuracy of the measuring device. It is known that radiation degradation of glass results in visible darkening, that is, increased absorption at longer wavelengths. UV/VIS spectra were taken before and after 31-day exposure and no change was found in the absolute transmittance or position of the absorption edge. The absorption coefficient of the glass was found to be about 0.06 cm^{-1} for the wavelength of emission peak, which is fine for waveguiding of visible light to the edges. The glass is therefore suitably compatible to soft tritium beta radiation to make a practical radioluminescent source.

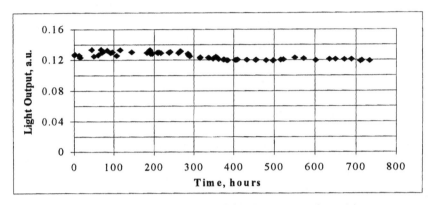

Figure 2. Relative light output of the glass exposed to tritium.

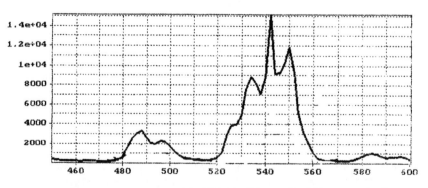

Figure 3. Emission spectrum of Ce/Gd-Tb glass under 270 nm excitation.

The main emission peak shown in Figure 3 is at 542 nm, which is typical for Tb^{3+} ion photoluminescence. It is assumed that the radioluminescent spectrum has a similar appearance to the photoluminescent spectrum, although the radioluminescent spectrum with tritium exposure was not directly measured in this work.

10 nA/cm^2 current was measured at the photocell. Due to source self-absorption, titanium tritide provides one-tenth the power flux of tritium gas used in conventional tritium bulbs for which hundreds of nA/cm^2 is typically obtained [9]. The open circuit voltage was 220 mV, giving a total power output of about 20 nanoWatts. For commercial uses, the product would require microwatts, obtained primarily by increasing the radioisotope concentration and scintillation glass surface area, among the other optimization strategies discussed.

Emission from the flat glass samples did not follow the Cosine Law. Angular distribution of the emission from the 5x1 mm edge of the 21x5x1 mm glass sample was studied. 5 mm was the edge around which rotation took place (see Figure 4).

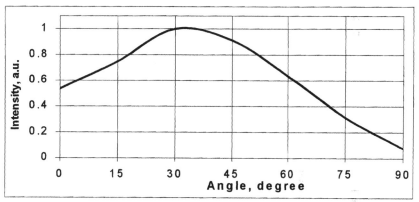

Figure 4. Angular distribution of light emitted from 5x1 mm edge
of the 21x5x1 mm glass sample.

The maximum intensity of emission is not at the zero angle but at 30-35 degrees because the beta absorption range in the glass is no more than 3 μm, while total thickness of glass waveguide was 1000 μm (1 mm). The brightest emissions emanated from very thin layers near the exposed surfaces. As the sample is rotated, more of one of the exposed surfaces (21x5 mm) is seen by the detector. However, the greater angle from the exposed surface reduces the emission accordingly to the cosine law. The measured dependence reflects a superposition of these two effects.

No mirrors were used in these experiments, since they would absorb part of the modest beta flux. Due to the total reflection effect, part of the light was nevertheless conveyed to the emitting edges of the glass. As one can see from Figure 5, light reflected within the solid angle 2ϕ escapes the waveguide, and only part of it is reflected back by titanium tritide sources deposited on steel substrates to provide additional light output.

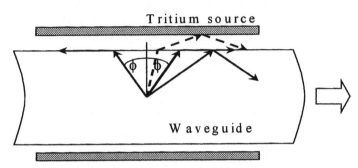

Figure 5. Scheme of light propagation in the waveguide.

However when light reaches the edge, total reflection plays a negative role. For better output, the emitting edge must be optically coupled to the photovoltaic which has a higher refraction index.

If the thickness of the waveguide is reduced to the thickness of beta range - several micrometers instead of current 1000 μm - the light output will increase two orders of magnitude. High refractive index coating of the waveguide with improved optical coupling to the photovoltaic will increase the light output proportionally to the ratio of the coated/uncoated area. These optimization steps are likely to provide $1\,\mu A/cm^2$ [9]. Assuming 3-4% efficiency of beta to light, and 30% of light to electricity conversion, 1% overall efficiency is possible. This is comparable to what has been achieved in the direct conversion approach [5].

CONCLUSIONS

We have discussed the advantages, feasibility and problems concerning indirect conversion of energy of radioactive decay into the electricity using scintillating glass. Improvement in the nuclear battery design will certainly facilitate MEMS technology and commercialization. Utilization of more powerful beta sources such as [147]Pm, and alpha sources, allows greatly increased power and longevity. Incorporation of the radioisotope directly into scintillating glass is worth consideration since it minimizes self-absorption and immobilizes the radioisotope at the same time. Radiation stability of the glass under the more energetic sources must be studied. This light concentrating RLS concept may be used not only for power generation but also in light sources for low intensity lightning as well as in microelectronics where self-sustained light sources are beneficial. This proof-of-concept work points the way to many exciting commercial opportunities for scintillation glass and radioisotopes.

ACKNOWLEDGMENTS

The authors gratefully acknowledge financial support provided by TRACE Photonics Inc. Assistance with electrical measurements by Drs. V.S. Kalinovsky and V.R. Larionov as well as the valuable advice of Drs. S. Deus, Yu.G. Shreter and Yu.T. Rebane is acknowledged.

REFERENCES
[1] R.G. Little, E.A. Burke, "Long life radioisotope-powered, voltaic-junction battery using radiation resistant materials", U.S. Pat. No. 5 440 187, Aug. 8, 1995

[2] L.C. Olsen, "Review of betavoltaic energy conversion", *Proc. of XII Space Photovoltaic Research and Technology Conf.*, 1992, p.256-267

[3] Frost & Sullivan MEMS Market Report, 2000

[4] E. Browne, and R.B. Firestone, Table of Radioactive Isotopes, John Wiley and Sons, New York, 1986

[5] S. Deus, "Alpha- and betavoltaic cells based on amorphous silicon", *Proc. of 16th EPSEC*, 2000, pp.60-64

[6] G.C. Rybicki, C.V. Aburto, R. Uribe, *Proc. of 25th IEEE Photovoltaic Specialists Conference*, 1996, p.93

[7] C. Bueno, R.A. Betz, M.H. Ellisman, G.G.Y. Fan "Hybrid luminescent device for imaging of ionizing and penetrating radiation", U.S. Pat. No. 5 594 253, May 20, 1996

[8] A.J. Bruce et al. "Er^{3+}-doped Soda-Lime Glasses and Planar Waveguide Devices"; pp.69-76 in *Synthesis and Application of Lanthanide-doped Materials*. Edited by B.G. Potter and A.J. Bruce. ACS, Westerville, 1995

[9] V.M. Andreev, "Advanced Betacell and Low-Intensity Photovoltaic Cells and Arrays Based on III-V Compounds", *Ioffe Institute Report*, Ioffe Institute. 2000

SYNTHESIS OF NANO-SIZED EUROPIUM DOPED YTTRIUM OXIDE

Burtrand I. Lee, Thomas S. Copeland, Amanda K. Elrod, and Jason Qi
School of Materials Sci. & Eng., Clemson Univ., Clemson, SC 29634 USA

ABSTRACT

Red light emitting Eu^{3+} doped Y_2O_3 phosphor nano-particles were synthesized by a sol-gel method combined with a furnace firing. All samples were prepared at various temperatures and times both with and without an organic surfactant in order to determine if the particle size is tailorable over a significant range. Firing it at 1100°C yielded full crystalline phase of Y_2O_3. The particle size, measured using dynamic light scattering, was between 36nm and 1 micron depending on the conditions. The microstructure of phosphor crystals characterized by scanning electron microscopy (SEM) showed spherically shaped particles. The SEM particle size measurements agreed well with the sizes determined by dynamic light scattering.

INTRODUCTION

In order to make flat panel displays (FPDs) more commercially viable, approaches to enhancing the luminescent properties of inorganic phosphors have been extensively investigated in recent years [1,2]. One objective has been to develop novel low-voltage cathodoluminescent (CL) phosphors with high efficiency and chemical stability under electron-beam bombardment in a high vacuum system for the next generation of field emission displays (FEDs) [3,4]. Even though they have satisfactory luminescent qualities, sulfide phosphors are unstable under electron beam exposure in a 10^{-7} torr vacuum, resulting in a chemical degradation of the phosphor layer and a detrimental effect on CL efficiency [5,6]. Non- sulfide phosphors are being researched in order to alleviate this instability. Yttrium oxide doped with trivalent europium ($Y_2O_3:Eu^{3+}$) is a common phosphor in cathode ray tubes. The Eu^{3+} ion is an attractive dopant because of its red fluorescence with high luminescent efficiency under UV light excitation resulting from the transition from the $^5D_0$ excited state to the ^7F states [7]. Solution approaches to $Y_2O_3:Eu^{3+}$ have included urea hydrolysis [8], sol-gel method carried out at near room temperature with low firing temperatures, different precursors, dopant amounts, and with added refluxing steps [9,10], and seeded hydrothermal methods [11]. In this paper we report a modified sol-gel method to produce submicron yttrium oxide powder in doped with europium ion.

EXPERIMENTAL

For this sol-gel procedure the $Y_2O_3:Eu^{3+}$, 16.90g of $Y(CH_3COO)_2 \cdot 4H_2O$ was stirred with 0.366g of $EuCl_3 \cdot 6H_2O$(4mol%) in deionized water for two hours at room temperature yielding a 1 molar solution. A Polyoxyethylene (20) sorbitan monooleate (Tween® 80, M.W.:1400 obtained from Aldrich, Madison, WI) solution was prepared by adding 1.18g (10 wt% of the finished products) of the polymer to water and adjusted to a pH = 10+ and a volume of 100 ml by using an aqueous ammonium hydroxide solution while being stirred. During the production of nano-scale yttria, Sharma, et al., [9,10] showed that an added surface modifier in an amount up to 10wt% like Tween could reduce the final particle size of their phosphor. Samples were also prepared in this fashion using ammonium polyacrylate (APA, M.W.:4000, obtained from Aldrich Chemicals) and KD-6 which is a copolymer of propylene glycol and methacrylic acid, available through Imperial chemicals, as the polymeric component and using 20% Tween to check for a point of diminishing return with respect to polymer addition. The solution was then dripped into the polymer solution at a rate of approximately 20 drops per minute under vigorous stirring. The resultant solution was then centrifuged to isolate the gel. This final gel product was then fired for two hours at temperatures varying from 800°C to 1100°C. $Y_2O_3:Eu^{3+}$ seeds were prepared by firing the above gel at 1100°C for 5 hours. Samples prepared included those with and without polymer (both APA and Tween), with 5wt% seeds and without seeding, and samples with varied maximum firing temperatures and times.

CHARACTERIZATION

Thermogravimetric analysis (TGA) and Differential Scanning Calorimetrty (DSC) measurements were performed using a NETZSCH STA 449C TGA unit in argon atmosphere from 50 °C to 1300 °C with a heating rate of 5 °C·min⁻¹. Scanning electron microscopy (SEM) micrographs were obtained from a Hitachi S-3500N Scanning Electron Microscope at 20 kV for pecimens prepared by coating with platinum on particles on carbon tape. Particle size measurements were carried out on a Brookhaven Dynamic Light Scatterer. X-ray diffractograms were obtained from Scintag XDS diffractometer from 2θ=20 to 80° with a scan step of 0.02°/Sec. Photoluminescence (PL) was determined by

LS-50B Perkin-Elmer Fluorescence Spectrophotometer with an excitation wavelength at 254nm.

RESULTS AND DISCUSSION

Figure 1 shows the particle size distribution of non-seeded and seeded $Y_2O_3:Eu^{3+}$ with 10wt% Tween sol-gel samples. They exhibit a mean diameter of 36nm with a broad distribution and 76nm with a narrower distribution respectively. Particle size is important for several reasons. Yoo and Lee [12] have shown that the optimum particle size can be a function of the anode voltage, electron penetration depth, and carrier generation rate. For larger particles at a low voltage, it was thought that electrons would collect at the surface creating a high electrostatic potential barrier at the surface preventing inhibiting further penetration. For a smaller particle, it should be easier for the electrons to penetrate past the high defect surface layer into the secondary conduction layer (containing significantly fewer defects). This is where the energy could be absorbed and transferred to the nearly, if not completely, defectless core for emission. Using Feldman's equations, Yoo and Lee [12] predicted that the optimum particle size for an anode voltage of 400V was 100nm. For CL, particle size becomes a large concern since the non-seeded sol-gel samples have a particle size well below their predicted optimal value and as the work voltage increases, so does the optimum particle size. Seeding promoted the particle size growth by aiding the nucleation but maintained narrower size distribution.

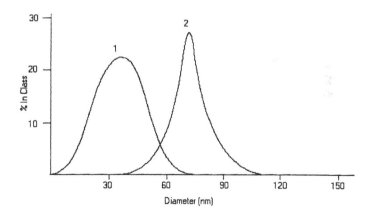

Figure 1. Particle size distribution of yttria/Eu powder: (1) a non-seeded with mean diameter of 36nm and (2) a seeded sample with a mean diameter of 76nm.

Table I lists the particle sizes of Y_2O_3:Eu^{3+} samples prepared under different conditions. This shows that the polymers added acted as the particle size growth inhibitors. Tween was the most effective inhibitor likely due to the greater adsorption on the Y_2O_3 surface. Tween is a nonionic surfactant and adsorbed on the negatively charged Y_2O_3 particle surface at the high pH environment. Other surfactants used being anionic, the adsorption on the negatively charged surface was more difficult.

Table I. Yttria/Eu powders prepared and the corresponding conditions and sizes

Sample #	Temperature	Time	Polymer	Seeds	Size
1	1100°C	2hours	10%Tween	None	36nm
2	1100°C	2hours	None	None	75nm
3	1100°C	2hours	20%Tween	None	52nm
4	1100°C	2hours	10%APA	None	108nm
5	1100°C	2hours	10%KD6	None	130nm
6	1100°C	2hours	10%Tween	5wt%	76nm
7	1100°C	2hours	10%APA	5wt%	111nm
8	450°C	2hours	10%Tween	None	50nm

Figure 2 shows the crystalline phase of Y_2O_3 after firing at 1100°C in air. No europium oxide phase could be seen in the structure at the doping level.

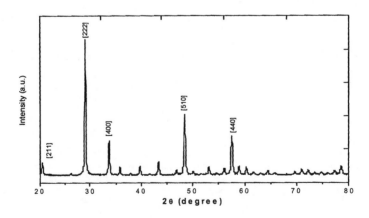

Figure 2. XRD pattern of Y_2O_3:Eu^{3+} prepared from acetate precursors and fired at 1100°C showing fully crystalline phase of Y_2O_3.

Figure 3 shows the SEM micrograph of a yttria sample with 20% Tween and no seeds fired at 1100°C. The particles have a roughly spherical shape, and are heavily agglomerated without being sintered. The individual particles have a diameter significantly less than 100nm and are fairly uniform.

Figure 3. SEM micrograph of yttria/Eu powder with 20% Tween/no seeds fired at 1100°C.

The amount of surface polymer and organic adsorbates on the yttria are shown indirectly in the TGA results in Figure 4. The total wt. loss of ~5% below 300°C

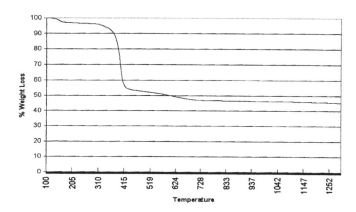

Figure 4. TGA of the gel derived from the precursor solution of yttrium acetate, europium acetate, and Tween.

is from adsorbed water in the surface polymer layer. The weight losses from 300°C to 430°C is from the major loss of chemically bound water and organics, acetate decomposition, and the decomposition of Tween®80. The weight loss for these species amounts to ~40%. The secondary loss of carbonaceous species at ~5% is shown in the range of 430°C to 700°C.

PL of red fluorescence at 614nm from 36nm Y_2O_3/Eu^{3+} phosphor is shown is shown in Figure 5. As we prepared the phosphors of different size, the PL intensity was shown to be inversely proportional to the particle size. This has been reported elsewhere [13]..

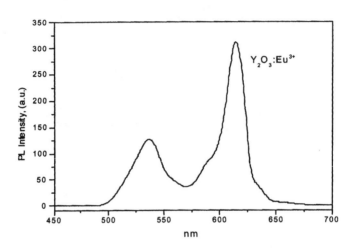

Figure 5 Photoluminescence of $Y_2O_3:Eu^{3+}$ excited at 254nm after fired at 1100°C.

CONCLUSIONS

Red light emitting $Y_2O_3:Eu^{3+}$ particles were produced using a sol-gel method. The addition of seeds increased the particle size significantly under all conditions. Both photoluminescence and cathodoluminescence were observed from these phosphor particles. The Tween was better than the APA in providing steric hindrance by providing a smaller micelle during particle formation and so decreased the final particle size from 108nm to 36nm.

REFERENCES

1) Extended Abstract, The Fifth International Conference on the Science and Technology of Display Phosphors, San Diego, November 1999.
2) P.D. Rack, A. Naman, P.H. Holloway, S.S. Sun, R.T. Tuenge, Mat. Res. Soc. Bull. 3 49-58 (1996).
3) S. Yang, C. Stoffers, F. Zhang, S.M. Jacobsen, B.K. Wagner, C.J.

Summers, Appl. Phys. Lett. **72** 158-160 (1998).

4) R.Y. Lee, F.L. Zhang, J. Penczek, B.K. Wagner, P.N. Yocom, C.J. Summers, J. Vac. Sci. Tech., **B 16** 855-857 (1998).

5) J. Ballato, J.S. Lewis, P.H. Holloway, Mat. Res. Soc. Bull. (1999) 51-53.

6) H.C. Swart, T.A. Trottier, J.S. Sebastian, S.L. Jones, P.H. Holloway, J. Appl. Phys. **83** 4578-4583 (1998).

7) J. Ballato, J.S. Lewis, P.H. Holloway, Mat. Res. Soc. Bull. (1999) 51-53.

8) Y. D. Jiang, Z. L. Wang, J. Mater. Res. **13** 2950-2955 (1998).

9) P.K. Sharma, M.H. Jilavi, D. Burgard, R. Nass, H. Schmidt, J. Am. Ceram. Soc. **81** 2732-2734 (1998).

10) P.K. Sharma, M.H. Jilavi, R. Nass, H. Schmidt, J. Mat. Sci. Lett. **17** 823-825 (1998).

11) S. Lu, B. I. Lee, T.S. Coperland, and C. Summer, J. Phys. Chem. Solid, **62**[4], 771-781 (2001).

12) J.S.Yoo, J.D.Lee, J. Appl. Phys. **81**(6) 15 March (1997).

13) T. Copeland, B.I. Lee, and A.K. Elrod, MRS Symp. Proc., Nov. 28-30, 2001, Boston, MA, in press., Materials Research Soc., Pittsburgh, PA.

UNUSUAL LONG-WAVELENGTH EXCITATION AND EMISSION IN Eu(II) AND Ce(III) DOPED M-Si-Al-O-N GLASSES

D. de Graaf, R. Metselaar, H.T. Hintzen and G. de With
Eindhoven University of Technology
PO Box 513, 5600MB
Eindhoven, The Netherlands

ABSTRACT

The luminescence properties of Ce^{3+} doped M-Si-Al-O-N glasses (M = Sc, Y, La, Gd) have been examined and the results compared with similar experiments on Eu^{2+}-doped Si-Al-O-N glasses. Both systems show a pronounced influence of the composition of the glass on the emission and excitation energy. As a function of the composition, and in particular the (Ce, Eu) content, the emission can be shifted over an interval exceeding 5 000 cm^{-1}, which is exceptionally large. It is further shown that these effects can produce, for Eu^{2+} and Ce^{3+}, extremely long wavelength emission. The nature of these effects is discussed.

INTRODUCTION

Si-Al-O-N glasses form during the liquid phase sintering of Si_3N_4 based ceramics when using Al_2O_3 and other metal oxides to promote the liquid phase formation [1]. These glassy intergranular phases dominate high temperature properties such as the creep resistance. For this reason some of the mechanical properties of M-Si-Al-O-N glasses have well been studied. These studies showed that the oxynitride glasses exhibit an enhanced mechanical, thermal and chemical durability as compared to oxide glasses [2,3]. This further increased the interest in the mechanical and chemical properties of these systems. However, as far as we know there has been no systematic study conducted towards the use of these systems as host materials for luminescent ions.

It is felt that the benefits of such a study are twofold. Firstly, the research towards Eu^{2+} and Ce^{3+} doped crystalline oxynitrides has yielded a number of materials with new luminescence properties [4]. Secondly, the 5d excited state of Ce^{3+} and Eu^{2+} is unshielded and thus sensitive to its local environment. It is our intention to use these ions as a local probe to study the changes in the glass structure as a function of the composition in order to enhance our understanding of the behaviour of modifier cations in sialon glasses. This information can aid the interpretation of the mechanical properties.

In a previous investigation the effect of Eu incorporation on the luminescence of (Y,Eu)-Si-Al-O-N glass has been studied. The luminescence characteristics were found to be peculiar. With increasing Eu content the band emission (characteristic for divalent europium) shifted towards longer wavelengths (Fig. 1).

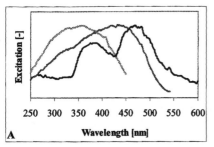 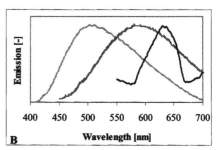

Fig. 1: Dependence of the excitation (A) and emission (B) spectra on the weighed-out composition of $Eu_xY_{35-x}Si_{45}Al_{20}O_{83}N_{17}$ glasses (x = 0.35 (■), 3.5 (■), 21.0 (■) [eq%]).

At high concentrations this leads to red emission (640 nm). Red emission from Eu^{2+} doped compounds has only been reported for a small number of materials. This red Eu^{2+} emission has consequently to be caused by a (for Eu^{2+}) very rare coordination.

It was decided to expand this investigation to the Ce^{3+}-doped sialon glasses. Ce^{3+} was chosen for a number of reasons. Firstly, Ce^{3+} has a similar 5d type excited state as Eu^{2+} [5]. Therefore a similar dependence of its luminescence on the coordination is expected. Secondly, Ce^{3+} has the same valence and a similar ionic radius as most of the other lanthanide ions, which makes it a better representative for this group of elements.

EXPERIMENTAL

Three types of compositional parameters were varied in this study. 1) Exchange of oxygen for nitrogen. 2) Exchange of cerium for aluminium. 3) Exchange of cerium for another trivalent modifier cation (scandium, yttrium, lanthanum, gadolinium). Weighed-out compositions are listed in Table I.

Table I. Chemical compositions of (M,Ce)-Si-Al-O-N glasses.

	Composition [eq%]*	x-values
1)	$Ce(27.6)\ Si(44.7)\ Al(27.6)\ O(100\text{-}x)\ N(x)$	2.9 4.4 7.0 11.4 13.3
2)	$Ce(x)\ Si(45)\ Al(55\text{-}x)\ O(83)\ N(17)$	22.2 24.5 26.4 30.3 34.7 36.4
3)	$Ce(x)\ Sc(35\text{-}x)\ Si(45)\ Al(20)\ O(83)\ N(17)$	2.6 8.5 19.3 35.0
	$Ce(x)\ Y(35\text{-}x)\ Si(45)\ Al(20)\ O(83)\ N(17)$	0.35 3.5 7.0 14.0 28.0 35.0
	$Ce(x)\ La(35\text{-}x)\ Si(45)\ Al(20)\ O(83)\ N(17)$	0.35 3.5 7.0 14.0 28.0 35.0
	$Ce(x)\ Gd(35\text{-}x)\ Si(45)\ Al(20)\ O(83)\ N(17)$	0.35 3.5 7.0 14.0 28.0 35.0

Starting materials (Sc_2O_3, Y_2O_3, La_2O_3, CeO_2, Gd_2O_3, Al_2O_3, SiO_2 and Si_3N_4) were weighed-out. Although added as CeO_2, Ce^{4+} was assumed to reduce to Ce^{3+} according to the reaction [6]:

$$12CeO_2 + Si_3N_4 \rightarrow 6Ce_2O_3 + 3SiO_2 + 2N_2 \uparrow \qquad (1)$$

This was accounted for by adjusting the SiO_2/Si_3N_4 ratio accordingly. Weighed-out powders were dispersed in isopropanol and mixed in a ball mill using Si_3N_4 balls. Thus obtained mixtures were dried and the powders pressed to pellets of ca. 1-2 g. The pellets were fired in an induction furnace using a BN powder lined, Mo crucible (N_2, furnace cooling). Melting times and current were adjusted per sample in order to obtain thoroughly molten glass droplets. The glasses were crushed prior to the optical characterisation. Fluorescence measurements were performed on a Perkin Elmer LS50B spectro-fluorometer equipped with a Xe flashlamp.

RESULTS AND DISCUSSION

All the glasses showed broadband emission, which is typical for Ce^{3+} containing materials. Changing nitrogen for oxygen has an insignificant effect on the emission and excitation characteristics of the glasses (Fig. 2). This means that nitrogen is not directly coordinated with cerium since replacement of oxygen for nitrogen in the first coordination shell of Ce^{3+} would have a substantial effect on the covalence and ligand-field strength of the site and thus on the luminescence characteristics. This nitrogen avoidance by cerium has also been reported in literature [6].

* Although weighed-out as CeO_2, compositions are calculated assuming cerium to be in the trivalent state.

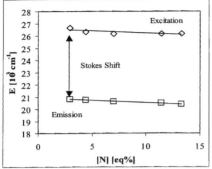

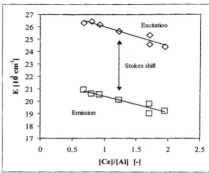

Fig. 2: Influence of nitrogen content on emission and excitation energy of Ce-sialon glasses.

Fig. 3: Dependence of the emission and excitation energy of Ce-sialon glasses on the [Ce]/[Al] ratio.

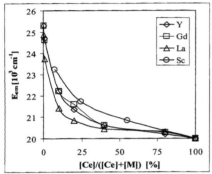

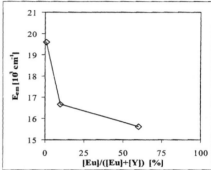

Fig. 4: Dependence of the emission energy of M-Ce-sialon glasses on the [Ce]/([Ce]+ [M]) ratio.

Fig. 5: Dependence of the emission energy of Y-Eu-sialon glasses on the [Eu]/([Eu]+ [Y]) ratio.

Substitution of cerium for aluminium, however, has a substantial effect on the luminescence characteristics. With increasing cerium content the emission as well as the excitation shift to lower energies (Fig. 3). The Stokes Shift not significantly affected by this change. Large changes can also be achieved by replacing cerium by another cation (e.g. Sc, Y, La, Gd). Again, with an increase of the cerium content the emission shifts to lower energies (Fig. 4). This general trend is observed for all four series, although a small difference between the series is visible at intermediate Ce contents. The magnitude of these shifts is remarkable. For the investigated compositions the emission shifts from 20 000 cm^{-1} (500 nm) to 25 300 cm^{-1} (395 nm).

Comparison of the data, which have been obtained for the cerium doped sialon glasses with our previous results on europium doped glasses show a very strong similarity (compare e.g. Figs. 4 and 5). The observed trends are very similar and the absolute changes of the emission energy are approximately the same. This shows that the effects, which are responsible for the emission changes in the cerium and in the europium doped glasses are the same.

The results show that the principal factor affecting the emission wavelength is related to the cerium/europium content. Energy transfer between the luminescent centres can explain this. Energy transfer lowers the emission energy and is directly related to the distance between luminescent centres and hence to their concentration. The average distance between the luminescent centres is expected to be sufficiently low for energy transfer to occur.

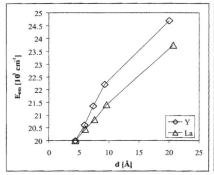 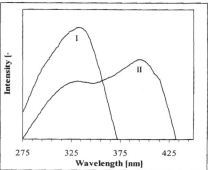

Fig. 6: Average Ce-Ce distance versus emission energy of M-Ce-sialon glasses (M = La,Y).

Fig. 7: Excitation spectra of $Ce_{0.35}Y_{34.65}Si_{45}Al_{20}O_{83}N_{17}$ (I) glass as compared to $Ce_{35}Si_{45}Al_{20}O_{83}N_{17}$ (II) glass.

Energy transfer is not the only factor affecting the emission. Plotting the emission wavelength against the average distance between the luminescent centres, which has been estimated from the molar volume of the glasses, shows that the host glass also influences the emission (fig. 6).

Stronger evidence for the presence of other effects is found in the excitation spectra, which show both in the case of cerium as in the case of europium a distinctive change of the ligand-field splitting of the excitation band with increasing cerium/europium content (Figs. 1A and 7). Therefore it is suggested that the coordination of cerium and europium change with the composition. This means that multiple sites are present in the glasses, which are preferentially occupied as a function of the composition. A likely source for the sites at high modifier contents is the formation of europium/cerium clusters or pairs. We therefore tentatively attribute these sites to clustering or pair formation.

CONCLUSIONS

This study shows that the emission energy of 5d type rare earth ions in sialon glasses is affected by at least two factors; one of them is energy transfer between the different sites in glasses. With increasing cerium/europium content the luminescent centres approach each other and the emission shifts to lower energies due to energy transfer. The magnitude of this shift is surprisingly large and leads to, for Eu^{2+} and Ce^{3+}, long wavelength emission. In the investigated range no influence of the nitrogen content on the emission was found. This means that similar effects are likely to occur in oxide glasses, although in oxide systems the effect will probably be less pronounced, since the introduction of nitrogen in most systems allows larger amounts of modifier cations to be incorporated in the glass.

A second effect is also taking place when incorporating a large amount of cerium/europium in the glass. In the excitation spectra a distinct change of the ligand-field splitting of the 5d-band is visible with an increase of the cerium/europium contents. We attribute the change of the coordination to pair formation or clustering. Further effort will be undertaken in the future to verify this hypothesis.

REFERENCES

[1]N. Hirosaki, A. Okada and K. Matoba," Sintering of Si_3N_4 with the Addition of Rare-Earth Oxides," *Journal of the American Ceramic Society,* **71** [3] C-144-C-147 (1988).

[2]R.E. Loehman, "Oxynitride glasses," pp.119-151 in *Treatise on materials science and technology,* Vol. 26, glass IV, Academic Press, London (1985).

[3]S.Sakka, "Structure, properties and application of oxynitride glasses," *Journal of Non-Crystalline Solids* **181**, pp. 215-244 (1995).

[4]J.W.H. van Krevel,"On new rare-earth doped M-Si-Al-O-N materials," *Ph.D. thesis,* Eindhoven University of Technology (2000).

[5]G. Blasse and A. Bril, "Characteristic luminescence," *Philips Technical Review* **31** [10], pp. 304-334 (1970).

[6]A. Díaz, S. Guillopé, P. Verdier, Y Laurent, A. López, J. Sambeth, A. Paúl and J.A. Odriozola, "XPS and DRIFTS study of Cerium in Ce-Si-Al-O-N glasses," *Materials Science Forum Vols.* **325-326,** pp. 283-288 (2000).

GAS-PHASE MODIFICATION OF THE DIRECT CURRENT ELECTROPHOSPHOR

V.D. Kupriyanov, N.A. Stepanova, B.M. Sinelnikov,
M.M. Sychov and V.G. Korsakov
St. Petersburg Institute of Technology
26 Moscow prospect St. Petersburg, Russia 198013

ABSTRACT
    Surface of ZnS direct current electrophosphor grains is usually modified from water solutions. Essential disadvantage of this method is considerable degradation of the modified electrophosphor brightness due to adsorption of the water vapours on its surface. We report in this paper results of direct current electrophosphor surface modification by the Atomic Layer Epitaxy method. Reactions between ZnS(Mn) electrophosphor and copper acetylacetonate vapours ware carried out at temperature of 453-493 K. Reaction products were studied by XPS, IR-spectroscopy and diffuse reflection spectroscopy. Degradation of brightness for the electrophosphor samples, modified by the Atomic Layer Epitaxy method is 25-30% less than that for samples obtained by treating in water solutions.

INTRODUCTION

    The peculiarity of the direct current powder electrophosphors is the presence of copper sulfide phase (Cu_xS) on the surface of grains. The heterojunctions $ZnS-Cu_xS$ representing the solid solutions are formed on the surface of the zinc sulphide direct current electrophosphors with composition ZnS(Mn). Cu_xS-phase is usually formed on the electrophosphor surface by modification with the use of the ion-exchange method from water solutions of mono- and bivalent copper salts. The usage of this method results in appearing of interface films of $Zn(OH)_2$ and CuS, decreasing the value of current, crossing the barrier, that causes decrease brightness of electroluminescence. Another disadvantage of the method is accelerated degradation of modified electrophosphor brightness due to absorption of the water vapours on electrophosphor surface with formation of the microhalvanic cells and distraction of ZnS-Cu$_x$S geterojunctions because of electrolysis caused by electric field. Gas-phase method of surface modification is free of such disadvantages.

EXPERIMENT

We report in this paper the results of ZnS(Mn) direct current electrophosphor surface modification by method Atomic Layer Epitaxy (ALE). Reactions between ZnS(Mn) direct current electrophosphor and copper acetylacetonate vapours were carried out at temperature of 453-493 K. To exclude water vapours and oxygen adsorption on the surface of electrophosphors the process of layering of Cu$_2$S-phase was carried in the reactor of flow type by using copper acetylacetonate (CAA) in the flow of the carrier gas. To obtain copper sulfide phase with high conductivity, the stoichiometric composition should approach the phase Cu$_2$S. So the process of modification of electrophosphor was carried out in reducing atmosphere (H$_2$:N$_2$ = 1:10). During the process of obtaining and storing of electrophosphors the film of ZnO formed. To remove that film and saturate the surface of electrophosphors with sulfer, they were preliminary treated in vapours of sulfur-containing compound (H$_2$S) in the temperature range 623-723 K. The reaction of sulfidation is shifted to the left, since the obtained H$_2$O vapours are being removed from the reaction zone, and hydrogen sulfide is delivers in excess.

XPS investigation[2] of the products of the interactions between electrophosphor surface and copper acetylacetonate vapours showed (tab. 1) that the value of the binding energy for Cu ^{2}p$_{3/2}$ level is 937.7 eV, that is lower than for copper acetylacetonate (932.9 eV), but higher than for Cu$_2$S (932.9 eV).

Table 1. The values of the binding energies of Cu ^{2}p$_{3/2}$ level

SAMPLE	The values of the binding energies, eV	Intensity, a.u.
Copper acetylacetonate	932.9	—
Electrophosphor after treatment with copper acetylacetonate	937.5	1.8
Electrophosphor after 1 cycle of treatment with CAA and H$_2$S	937.5	1.7
Electrophosphor after 4 cycles of treatment with CAA and H$_2$S	933.1	5.1

Optoelectronic Materials and Technology

$$\begin{array}{ccc}
 & CH_3 & CH_3 \\
 & | & | \\
- Zn - S & O - C = CH - C = O & \\
 & \diagdown \diagup & \\
 & Cu & \\
 & \diagup \diagdown & \\
- Zn - S & O - C = CH - C = O & \\
 & | & | \\
 & CH_3 & CH_3
\end{array}$$

Figure 1. Complex of copper acetylacetonate with ZnS grain surface.

The infrared spectra of direct current electrophosphor samples modified with copper acetylacetonate have absorption bands with maxima at 1370, 1470, 2850 and 2960 cm^{-1}, which may be related with presence of organic groupings. It may be supposed, that at the interaction of copper acetylacetonate vapours with the electrophosphor surface in the given interval of temperatures the breaking of helate bounds takes place with formation of surface complex (fig. 1).

Increase of the temperature in the process of the hydrogen sulfide treatment in the flow of carrier gas to 513-533 K results in decomposition of the surface complex and releasing of volatile organic compounds (acetone, carbon dioxide, carbon monoxide), removed from reaction zone by carrier gas. At this stage in IR spectra of samples the absorption bands, characteristic for organic groupings, disappeared. The binding energy of the Cu $^2p_{3/2}$ level becomes equal to 933.5 eV according to the XPS data.

XPS data shows, that as the number of cycles of electrophosphor treatment by copper acetylacetonate and hydrogen sulfide vapours increases, the magnitude of binding energy of the Cu $^2p_{3/2}$ level approaches the magnitude of the binding energy, characteristic of bulk copper sulfide. The intensity of the peak with the increase of the number of treatment cycles becomes larger, that shows the increase of copper content in the samples.

The optimal specific resistance ρ_v of direct current electrophosphors according to paper[3] lays in the range of 40-70 Ohm*cm. After 2 cycles of treatment by copper acetylacetonate and hydrogen sulfide vapours resistance of the samples was 65 Ohm*cm, and after 4 cycles of treatment – 43 Ohm*cm.

The diffuse reflection spectroscopy[4] data for electrophosphors, modified by atomic layer epitaxy method and nonmodified are shown on fig. 2.

Table 2. Characteristics of electrophosphors

Method of modification	Number of treatment cycles	Brightness, a.u.	Brightness degradation, a.u.	ρ_v, Ohm*cm	Size of particles, μm
ALE	1	110	98	90	3
-//-	2	165	83	65	3.1
-//-	3	167	80	51	3.3
-//-	4	173	75	43	3.4
From water solutions	—	100	100	170	5.2

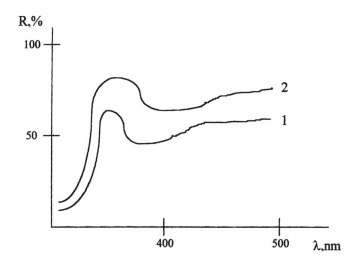

Figure 2. Diffuse reflection spectra: 1 – nonmodified electrophosphor;
2 – ALE modified electrophosphor

The optical and technical characteristics of electrophosphors modified by 1-4 cycles of treatment by copper acetylacetonate and hydrogen sulfide vapours, and those for electrophosphor modified from water solutions[5] are shown in tab. 2. The study of the parameters of the electrophosphors was performed in the dismountable electroluminescent cell at imposed voltage $U = 100$ V. As it follows from the data of the tab. 2, direct current electrophosphors, modified by Atomic Layer Epitaxy method with treatment by copper acetylacetonate and hydrogen sulfide vapours enables to increase the value of the brightness and its stability, and also to prevent aggregation of the electrophosphor grains in the process of modification. All these factors permit to improve the characteristics of the electroluminescent devices (breakdown voltage, brightness, stability of characteristics and reliability).

CONCLUSIONS

The surface modification of ZnS(Mn) direct current electrophosphors by Atomic Layer Epitaxy method with treatment by copper acetylacetonate and hydrogen sulfide vapours permitted to obtain samples with 60-70 % increased brightness and 25-30 % decreased brightness degradation comparing to samples obtained by treatment from water solutions.

REFERENCES

[1] Yu.N. Verevkin, "Degradation Processes in Electroluminescence of Solids", Nauka, Leningrad, 1983. p.122

[2] V.I. Nefedov, "Roentgen-electron Spectra of Chemical Compounds", Khimia, Moscow, 1984. p.250

[3] B.N.Korolko, A.Yu. Semchuk, "Powder Electroluminescent Direct Current Devices", IFANUSSR, Kiev, 1984. p.230

[4] V.N. Pak, N.G. Ventov, "Diffusive Reflectance Spectra", *Journal of Applyed Chemistry*, [3] 1678 (1974).

[5] I.K. Vereschagin, "Electroluminescence of Crystals", Nauka, Moscow. 1974, p.350

Novel Synthesis of Amorphous and Semiconducting Optoelectronics

LOCAL STRUCTURE AND RAMAN VIBRATIONAL SPECTRA OF DOPED TELLURITE GLASSES

H.M.Moawad[2], J.Toulouse[1]*, H.Jain[2], O.Latinovic[1], A.R.Kortan[3]
1-Physics Department, Lewis Laboratory.,
2-Materials Science and Engineering Department,
Whitaker Laboratory,
Lehigh University, Bethlehem, PA,18015
3-Lucent Technologies/Bell Laboratories,
600 Mountain Avenue, Murray Hill, NJ 07974

July 20, 2001

Abstract

Tellurite glasses possess properties of great interest for optical applications, the primary ones being high transparency and high indices of refraction, both linear and nonlinear. The best optical TeO_2 glasses are those containing Na_2O and ZnO. Using the combination of X-ray photoelectron and Raman spectroscopies, we have investigated a series of Na_2O-ZnO-TeO_2 glasses of different compositions. This combination allows the determination of the "molecular" units present in the glass as well as their arrangement in the network. These results can then be correlated with the optical properties.

1 INTRODUCTION

Tellurite glasses are complex in that the building blocks or structural units composing them change with the introduction of dopants/modifiers or with the

---

*corresponding author

---

processing method used. However, they are simpler than other complex glasses, such as borates, by the limited number of such units.

Tellurite glasses have attracted of a great deal of attention recently, because their properties are of interest for optical applications. The most promising applications are in rare-earth-doped or Raman fiber lasers and amplifiers. For these applications, the most relevant properties include, linear and nonlinear indices of refraction, transmissivity in the infrared, dopant solubility, phonon frequencies and, last but not least, chemical stability. At a fundamental level, the most central properties are the polarizability and the electron-phonon coupling. The primary advantage of TeO_2 glass is a much higher index of refraction than that of silica (2.3-2.4 vs 1.4-1.5) with a comparable and excellent chemical stability. TeO_2 glass also offers a high solubility for rare earth ions, erbium or thullium in particular, and relatively low phonon frequencies. Finally, the multivalent states of Te and the low bonding strength of the Te-O bond provide compositional flexibility while low glass transition temperatures ensure ease of fabrication.

X-ray Photoelectron (XPS) and Raman spectroscopies constitute a particularly powerful combination of techniques for a fundamental study of the structural origin of the vibrational dynamics of TeO_2 glasses. XPS is a species-selective technique that helps identify the structural units present on the basis of the binding energy of specific electrons. Raman spectroscopy can also help identify structural units on the basis of their high frequency internal vibrational modes but, in addition, can provide information relative to the network topology on the basis of lower frequency intermolecular or collective modes. Moreover, the Raman scattering amplitude, or the Raman scattering cross-section, is obviously the quantity of interest for applications such as Raman lasers and amplifiers. The tellurite glasses being presently considered for fiber lasers and amplifiers must be highly transparent, while still doped with relatively high concentrations of erbium or thullium. At present, the compositions that seem to best meet these two conditions include Na_2O and ZnO. The binary TeO_2 glasses with either Na or Zn also provide a wide glass forming region, contributing to the compositional flexibility mentioned above, as well as an excellent transparency. Therefore, we have undertaken the XPS-Raman study of a series of Na_2O-ZnO-TeO_2 glasses.

2 EXPERIMENTAL RESULTS AND DISCUSSION

The tellurite glasses were prepared from Purotronic-grade materials, melted in gold crucibles at 800-1000°C, purified in a chlorine-oxygen atmosphere and, finally, quenched from 800°C down to T_g on a gold surface. Three groups of three compositions each were prepared: a binary ZnO-TeO_2, a ternary with 5% Na_2O and a ternary with 10% Na_2O (molar concentrations). The compositions studied are listed in Table I below.

Table I. Compositions of the glasses studied

Samples	1	2	3	4	5	6	7	8	9
TeO_2	80%	72.5%	65%	75%	67.5%	60%	75%	67.5%	60%
ZnO	20%	27.5%	35%	20%	27.5%	35%	15%	22.5%	30%
Na_2O				5%	5%	5%	10%	10%	10%

2.1 X-Ray Photoelectron Spectroscopy

X-Ray Photoelectron spectroscopy measures the flux of electrons emitted from the sample as a function of their energy or, equivalently, as a function of their binding energy. It is a species-selective technique in the sense that the binding energy of the emitted electrons depends upon the particular atom from which they are emitted. In the present experiments, we have probed the 1s electrons emitted from oxygen atoms and the $3d_{5/2}$ electrons emitted from tellurium atoms. Fig. 1 shows a typical 1s oxygen spectrum, this one made up of two peaks, one from bridging oxygen (BO) and the other from non-bridging oxygen (NBO) atoms. The simplicity of the oxygen spectrum indicates that the binding energy of the 1s electron is primarily influenced by the presence of one or two nearest neighbor tellurium atoms and, only to a much smaller extent, by the presence of other oxygen atoms nearby. These peaks are extracted in a two-peak fit of the experimental spectrum using a Voigt profile. Both the Voigt profile and the binding energy for a particular peak are kept constant for the fits of all spectra. The identification of a particular peak is made on the basis of the binding energy of the electron emitted by a particular atom, which is inversely proportional to the effective charge of that atom. This effective charge itself depends upon the particular structural unit in which this atom is found and can be calculated using Pauling's fractional ionicity formula[1]:

$$Q_A = \sum_B b_B I_{AB}$$

where I_{AB} is the fractional ionicity of the AB bond, defined as:

$$I_{AB} = 1 - (R_A/M_A) \exp[-0.25(\lambda_A - \lambda_B)^2]$$

in which R_A is the average valence, M_A, the coordination number and λ, Pauling's electronegativity.

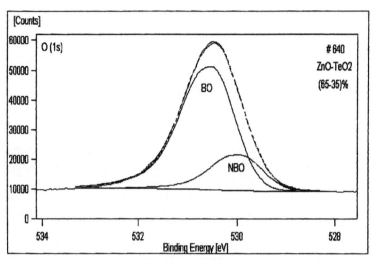

Figure 1 XPS spectrum of the 1s electron of oxygen

The spectrum of the $3d_{5/2}$ valence electron of tellurium is shown in Fig.2. It is more complicated than that of the oxygen spectrum because there exist four possible structural Te units, respectively labeled TeO_4, TeO_3, TeO_{3+1} and Te_3O_8.

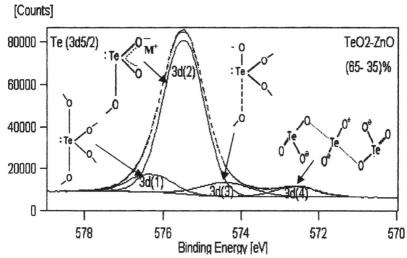

Figure 2 Tellurium $3d_{5/2}$ XPS spectrum

. It is noteworthy that excellent fits could be obtained with very similar binding energy parameters for a given-peak in all the spectra, indicating that the units present in the different glasses are very well defined. Greater widths indicate a broader distribution of local network environments. The fraction of each type of unit can be estimated from the integrated intensity of the corresponding peak. The fractions of NBO's and of the different Te units are presented in Figs. 3 and 4 as a function of the TeO_2 concentration. In Fig.4, the fraction of TeO_3 and TeO_{3+1} units, taken together as N_3, relative to TeO_4 units. As should be expected, a decrease in TeO_2, or a corresponding increase in Na_2O and ZnO, results in an increase in the fraction of NBO's and in N_3. This trend intensifies at lower concentrations of TeO_2.

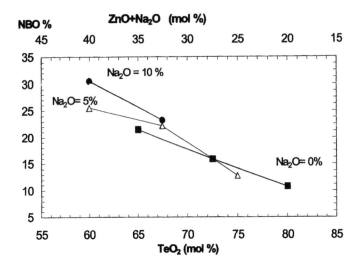

Figure 3 NBO dependence on TeO₂ concentration

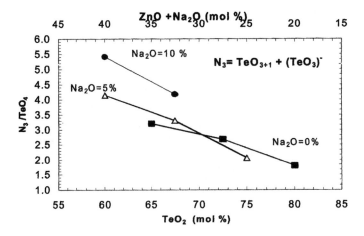

Figure 4 N3 dependence on TeO₂ concentration

Optoelectronic Materials and Technology

However, unexpectedly in the $75\%TeO_2$-$5\%Na_2O$-$20\%ZnO$ ternary glass, both the numbers of NBO's and N_3 fall below their value in the TeO_2-ZnO binary. It thus seems that, in this particular compositional range, the ternary glass possesses a higher connectivity than the binary glass. This conclusion is further confirmed by a more detailed examination of the XPS results, which indicates that a greater fraction of the TeO_4 units subsists in the ternary $5\%Na_2O$ glasses than in both the binary or the ternary $10\%Na_2O$ ones. Concurrently, the ternary $5\%Na_2O$ glasses contain fewer $(TeO_3)^-$ units but more TeO_{3+1} units. This latter result also suggests a more connected network in the presence of Zn than in simple binary alkali tellurites. Another interesting result is the appearance of a fourth peak in the Te-$3d_{5/2}$ spectrum of the binary 65%- 35% glass.(fig.2). This fourth peak, at a lower binding energy, can be attributed to a Te_3O_8 structural unit, in which tellurium has a higher effective charge. It is interesting to note that the β-form of crystalline tellurite is actually made up of such units. To the first degree, the conclusion of the XPS part of the study is that, for lower TeO_2 concentrations, Na and Zn in tellurite glasses act as modifiers and contribute to the conversion of TeO_4 primarily to TeO_3 units. However, for higher TeO_2 concentrations, Na and Zn taken together act more as glass formers, giving a more highly connected glass than the binary glasses. At high concentrations, Zn itself tends to transform the network by creating Te_3O_8 structural units that can link with one another in chains through the same double Te-O-Te bridges found in the β-form of crystalline tellurite.

2.2 Vibrational Raman Spectroscopy

Raman Spectroscopy is complementary to XPS and a valuable probe in two respects. Raman is the dynamic complementary of XPS because it helps identify the structural units present in the glass through their vibrational modes. These can be internal vibrational modes of a particular unit but also collective modes of several units (e.g. torsional modes of a chain). Thus, Raman is capable of providing also information about the network and its connectivity. Secondly, the Raman scattering intensity is directly proportional to the strength of the electron-phonon coupling, and therefore contains information, not only on the displacement field, Q, of a particular vibration, but also on the magnitude of the electronic polarizability, α, which can be expanded as:

$$\alpha_{ij} = \alpha_0 + \sum_k \left(\frac{\partial \alpha_{i,j}}{\partial Q_k}\right)_0 Q_k$$

A typical Raman spectrum of tellurite glasses is shown in Fig.5. It is composed of three groups of peaks: 550-850cm^{-1}, 250-550 cm^{-1} and 50-250cm^{-1}. The vibrations of single or isolated bonds and the vibrations of bridges between individual units are visible in the two higher ranges.

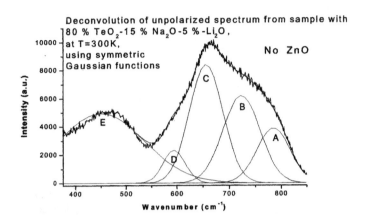

Figure 5 Typical Raman spectrum of TeO2 glasses

Sekiya et al.[2],[3] have successfully shown that the highest frequency group could be deconvoluted into four peaks, A at 780 cm^{-1}, B at 720 cm^{-1}, C at 665 cm^{-1} and D at 611 cm^{-1}. The assignment of these peaks has been made on the basis of corresponding peaks observed in the spectrum of crystalline paratellurite (α-form).[4] The two highest frequency peaks, A and B, can clearly be assigned to stretching modes of (Te-O)$^{-}$ bonds containing an NBO and, as such, are not present in pure TeO$_2$ glass but appear with the introduction of dopants/network modifiers. The next two peaks, C and D, are strong in pure TeO$_2$ glass and, as such, are characteristic of TeO$_4$ units. They have been attributed to antisymmetric stretching modes of Te- O-Te linkages, made up of two unequivalent Te-O bonds (one equatorial oxygen and the other axial). The peak(s) at intermediate frequencies, E at 450 cm^{-1}, has not been unequivocally assigned, being attributed to either symmetric stretching or bending modes of Te-O-Te linkages. In paratellurite, a corresponding mode has been assigned to the bending of identical bonds, O_{ax}-Te-O_{ax} and O_{eq}-Te-O_{eq}.[4] It is also interesting to note that, in silica SiO$_2$, a peak at 440 cm^{-1} known as "the broad band" is clearly related to the bending motion of Si-O-Si bridges.[5] The assignments

Optoelectronic Materials and Technology

of the C peak at 665 cm^{-1} and of the E peak at 450 cm^{-1} to Te-O-Te linkages means that the disappearance of these two peaks upon introduction of modifiers in the glass network provides a clear sign of the loss of connectivity. Finally, the low frequency peaks, 50-250 cm^{-1}, have been assigned to bending and torsional modes. We shall present below supporting evidence for the torsional nature of some of these modes. The lowest mode, at $\sim$40 cm^{-1}, is the well-known Boson peak, a collective mode most likely involving acoustic phonons, and which is present in all glasses.

We now compare the Raman spectra of glasses with different compositions and attempt to extract information on the respective and combined effects of Na and Zn as modifiers or glass formers in telluride glasses. We first examine the spectra of the TeO$_2$-ZnO binary glasses shown in Fig.6. Those spectra as well as all the other presented in this paper have been normalized to the height of the Boson peak for comparison. This choice of reference is the most appropriate because the Boson peak, being at the lowest frequency, also corresponds to the most extended of the modes, and its height is therefore expected to be least sensitive to the particular composition of the glass. This assumption is supported by the fact that a Boson peak is observed in all glasses or amorphous systems, in the same frequency range although not always with the same magnitude[5].

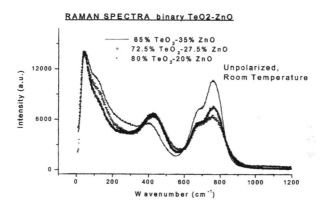

Figure 6 Raman spectra of binary ZnO-TeO2 glasses

The spectra of the TeO$_2$-ZnO binary glass exhibit the three groups of peaks described above. At the highest frequencies, an increase in the relative concentration of ZnO results in an increase of the A and B over the C and D peaks.

According to the above identification of these peaks, such an increase signifies the formation of (Te-O)⁻ bonds. In alkali tellurites, one expects the latter bonds to appear when TeO_4 units convert to $(TeO_3)^-$ units. However, in the present spectra, the C and D peaks also increase in magnitude with increasing Zn concentration and this is particularly visible in the top trace for the 35%ZnO glass. One tentative explanation is that the presence of Zn modifies the network in such a way as to introduce NBO's in (Te-O)⁻ bonds as well as Te-O-Te linkages, as in the perfectly connected TeO_2 glass. In fact, we reported above, in the XPS spectra of the same glass, a fourth peak corresponding to new Te_3O_8 structural units. Each such unit does contain four NBO's (see model in fig.2) and, in addition, two asymmetric Te-O-Te linkages, which might explain why both pairs, A-B and C-D, increase simultaneously. In the same spectrum of the 35%ZnO binary glass, the E peak decreases and additional intensity appears between 100 and 300 cm⁻¹. This observation is quite consistent with our previous interpretation; if new Te_3O_8 structural units are formed, that are linked in chains, the bending modes will necessarily be strongly affected. Alternatively, chains will give rise to new low frequency modes of torsion, particularly visible as a shoulder to the Boson peak at ∼115 cm⁻¹. Comparing the spectra of the 35%ZnO with those of the two lower Zn concentrations, one sees that the latter shows an increase in the A and B peaks but only little change in the C and D or in the E peaks. This suggests that, at lower concentrations, Zn is overall a network modifier rather than a network former. Nonetheless, the appearance of some additional intensity at lower frequencies already suggests the formation of isolated Te_3O_8 units

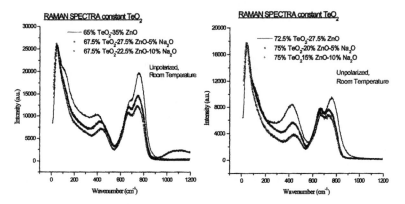

Figure 7 Raman spectra of ternary glasses with constant TeO2, varying Na/Zn

Optoelectronic Materials and Technology

·We next examine, in Fig.7, the relative influence of Na and Zn on telluride glasses for two different concentrations of TeO_2, 65-67.5% and 72.5-75%. Comparing the two sets of spectra, it appears that the influence of the Na/Zn ratio is greater at the highest and lowest frequencies for a higher concentration of TeO_2 but greater at intermediate frequencies for a lower concentration of TeO_2. Looking first at the highest frequencies, Na is seen to prevent the transfer of scattering intensity from the C-D to the A-B peaks, or, conversely, Zn enhances this transfer. Accordingly, at the lowest frequencies, the torsional modes are much less present with the substitution of Na for Zn. Thus, Na appears to preserve the overall 3-dimensional connectivity or isotropic topology of the network as opposed to Zn, which favors a transformation to a more one-dimensional connectivity or anistotropic chainlike topology, with the creation of Te_3O_8 units. At intermediate frequencies, the E peak starts more intense but decreases more rapidly with increasing Na/Zn ratio for a higher TeO_2 concentration. This evolution indicates that the Na-Zn substitution is more effective in severing the Te-O-Te linkages for higher TeO_2 concentrations. Such a result is quite consistent with the creation of Te_3O_8 units, or even chains of these, induced by Zn and the concurrent loss of the original Te-O-Te linkages between TeO_4 units. Finally, one can notice that, for a higher TeO_2 concentration and a higher Na/Zn ratio, the low frequency shoulder around 115 cm^{-1} is much less visible. Thus again, even in the presence of individual Te_3O_8 units, Na prevents the formation of the chain-like structure favored by Zn, preserving instead the isotropic topology of the original TeO_2 glass

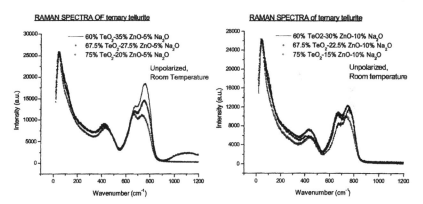

Figure 8 Raman spectra of ternary glasses with constant NaO, varying Zn/Te

·We finally compare, in Fig.8, the effect of the ZnO/ TeO$_2$ ratio on the Raman spectrum, for identical concentrations of Na$_2$O. One notices a significant change upon going from 5% to 10%Na$_2$O. Focusing on the high frequency group of peaks, the spectra of the 5% Na$_2$O glasses resemble those of the TeO$_2$ -ZnO binary glass and are therefore dominated by the influence of Zn, which suggests the presence of Te$_3$O$_8$ units. At the same time, very little change in intensity is observed in the low frequency group, which indicates the absence of a chain-like structure responsible for the torsional modes. Thus, we conclude that the 5% Na$_2$O glasses only contain individual Te$_3$O$_8$ units embedded in a 3D-connected network. In fact, the slight increase of the E peak with increasing ZnO concentration in the 5%Na$_2$O series suggests that the substitution of Zn for Te even slightly enhances the number of Te-O-Te linkages and, consequently, the overall connectivity of the glass. By contrast, the spectra of the10%Na$_2$O glasses exhibit a much weaker dependence on the ZnO/ TeO$_2$ ratio and must be dominated by the presence of Na. Here, increasing the ZnO/ TeO$_2$ ratio results in a loss of intensity in the C and D peaks and a gain of intensity in the A and B peak, as expected when TeO$_4$ units are converted into (TeO$_3$)$^-$ and TeO$_{3+1}$ units. From the evolution of the E peak intensity with the ZnO/ TeO$_2$ ratio in these glasses, it appears that replacing Te by Zn does not affect the Te-O-Te linkages significantly. Initially, this even appears to raise their number slightly. The limited increase in the 115 cm^{-1} shoulder with increasing ZnO concentration also indicates that the topology of the network, is only slightly affected by the presence of Zn. Comparing these two series, it is also worth noting that the Raman intensities of the 5%Na$_2$O glasses are significantly higher than those of other compositions, particularly in the high frequency region (peaks A-D). The introduction of Te$_3$O$_8$ units in these glasses must therefore increase the Raman cross-section, or equivalently the modulation amplitude of the polarizability by phonons or vibrations. Since the Raman cross-section is proportional to the imaginary part of a third order nonlinear susceptibility, this also suggest that the real part of this susceptibility, $\chi^{(3)}$, may also be enhanced in these glasses.

3 CONCLUSIONS

In the present study, we have reported and compared experimental results obtained from X-ray photoelectron and Raman spectroscopies. The first technique has allowed the identification of the basic (molecular) units forming in Na$_2$O-ZnO-TeO$_2$ glasses, and the second one, the identification of the same units as well as of their spatial arrangement in the network. The latter information has been obtained from the observation of the vibrations of intra- and intermolecular bonds. The partial conclusions reached in the two parts of this study are in excellent agreement. Several major points can be made:

·The main effect of Na and Zn in TeO_2 glasses ins to break up Te-O-Te linkages between TeO_4 units and provoke their conversion into TeO_3 and TeO_{3+1} units containing at least one non-bridging oxygen (NBO).

·Na tends to decrease the connectivity while preserving the 3-dimensional topology (or the isotropic character) of the glass network.

·By contrast, Zn rebuilds the network, inducing the formation of Te_3O_8 units that can be linked in chains. It is worth noting that these units and chains are present in the β-form of crystalline tellurite, in which they are also linked transversally, producing of a layer structure.

·Na tends to prevent the formation of chains in ZnO-TeO_2.

·5%NaO-20%ZnO-75%TeO_2 appears to fall in a special concentration range in which the Zn content is high enough so as to induce the formation of Te_3O_8 units while the Na content is sufficient to prevent the formation of chains but not so high as to reduce the connectivity. The Te_3O_8 units are then embedded in a 3-dimensional network that appears to be more highly connected than would be in either of the two corresponding binary glasses.

This work was supported by a grant from the National Science Foundation, NSF-DMR-9974031. Special thanks to N. Kopylov for the glass fabrication and to A.Miller for help with the XPS measurements.

References

[1] R.K.Brow and C.G.Pantano, J.Am.Ceram.Soc., 69, 314 (1986)

[2] T.Sekiya, N.Mochida, A.Ohtsuka and M.Tonokawa, J.Non-Cryst.Solids 144, 128 (1992)

[3] T.Sekiya, N.Mochida, A.Ohtsuka, J.Non-Cryst.Solids, 168, 106 (1994)

[4] T.Sekiya, N.Mochida, A.Ohtsuka and M.Tonokawa, Journal of the Ceramic Society of Japan, 97,1435 (1989)

[5] C.M.McIntosh, J.Toulouse, P.Tick, J.Non-Cryst.Solids, 222, 335 (1997)

EFFECTS OF STARTING COMPOSITIONS ON THE PHASE EQUILIBRIUM IN HYDROTHERMAL SYNTHESIS OF Zn_2SiO_4:Mn^{2+}

Chulsoo Yoon and Shinhoo Kang,
School of Materials Sci. and Eng.
Seoul National University
Shinlim-dong San 56-1, Kwanak-gu
Seoul 151-742, Korea

ABSTRACT

The hydrothermal technique is one of the wet-chemical methods to produce phosphor for display applications. In this study Mn-doped Zn_2SiO_4 phosphors were hydrothermally synthesized using ZnO and SiO_2. When the composition of the raw materials was consistent with the stoichiometry of Zn_2SiO_4, hemimorphite $(Zn_4Si_2O_7(OH)_2 \cdot H_2O)$ phase formed under the condition of low temperature and pressure. The phase transition from $Zn_4Si_2O_7(OH)_2 \cdot H_2O$ to Zn_2SiO_4 occurred at 350~360°C under an external pressure of 140MPa. However, it was found that the Zn_2SiO_4 phase could be precipitated at low temperature by adjusting the composition of raw materials. Investigation of actual doping concentrations of Mn in $Zn_{2-x}Mn_xSiO_4$ revealed that about 10% of initial Mn content were doped in Zn_2SiO_4 lattice. The photoluminescence of the synthesized phosphor showed maximum brightness at x=0.02, which is lower concentration than reported elsewhere. This was due to the increase in residual SiO_2 at higher x value.

INTRODUCTION

Due to its high quantum-efficiency in the vacuum ultraviolet (VUV) region, Zn_2SiO_4:Mn^{2+} is employed as a green phosphor for color plasma display panels (PDP).[1] For the preparation of these materials solid state reaction has been adopted as a common route of synthesis.[2-4] However, the presence of large agglomerates and particles of irregular morphology was unavoidable due to the high calcination temperature (1200~1300℃) in this process.[5]

The hydrothermal process is a cost-effective method that can produce ceramic powders having a narrow size-distribution, regular shape, and high crystallinity.[6,7] Efforts have been directed at preparing Zn_2SiO_4:Mn^{2+} using a hydrothermal technique by some researchers and successful syntheses of these materials were reported.[8-10] Recently, the mechanism of phase formation and luminous characteristics of $Zn_2SiO_4:Mn^{2+}$ phosphors synthesized hydrothermally were reported.[11]

In this study, $Zn_2SiO_4:Mn^{2+}$ phosphor were synthesized hydrothermally and the evolution of equilibrium phases at different temperatures and compositions were investigated. The key parameters for the hydrothermal synthesis of $Zn_2SiO_4:Mn^{2+}$ were surveyed. In addition, Mn-doping phenomena under hydrothermal conditions were studied and relationships between the amount of added Mn and the actual Mn doping concentration after synthesis were investigated. The photoluminescence characteristics of the synthesized $Zn_2SiO_4:Mn^{2+}$ phosphors were also estimated.

EXPERIMENTAL

Starting materials used for host materials were ZnO (Cerac, Milwaukee, WI) and SiO_2 (Sigma, St. Louis, MO). $MnCl_2 \cdot 4H_2O$ (Aldrich Chem., Milwaukee, WI) was used as a raw material for activator. The mixtures of raw materials were ball-milled for 72h and dried. After mixing, the powders were dispersed in distilled water and put into Teflon-coated autoclave. Hydrothermal synthesis was performed at 150~200 ℃ under autogenous pressure.

After filtering and washing, green emitting $Zn_2SiO_4:Mn^{2+}$ phosphor particles were obtained. These powders were characterized by x-ray diffraction (XRD), scanning electron microscopy (SEM), inductively coupled plasma (ICP), and photoluminescence (PL) analysis.

RESULTS AND DISCUSSION

As shown in Fig. 1, $Zn_4Si_2O_7(OH)_2 \cdot H_2O$ forms as an equilibrium phase when the $ZnO-SiO_2-H_2O$ system was hydrothermally heat-treated at 200 ℃ under autogenous pressure. The formation of $Zn_4Si_2O_7(OH)_2 \cdot H_2O$ phase occurs through the dissolution-reprecipitation after heterogeneous nucleation of dissolved raw materials on ZnO particles.[11] In this case, the molar ratio of ZnO/SiO_2 in the starting powder mixture was adjusted to the stoichiometry of Zn_2SiO_4. The phase transition from $Zn_4Si_2O_7(OH)_2 \cdot H_2O$ to Zn_2SiO_4 was accomplished when the synthesis temperature reached 360 ℃.

In order to increase the yield and decrease the cost of powder production of $Zn_2SiO_4:Mn^{2+}$ phosphor in hydrothermal process, it is prerequisite that the synthesis temperature and pressure should be lowered. When the composition of raw materials was modified, the formation of Zn_2SiO_4 phase occurred at 200 ℃ under autogenous pressure as shown in Fig. 2. The molar ratio of $ZnO:SiO_2$ was adjusted to 1.95:1 whereas the composition of starting powder was $ZnO:SiO_2=2:1$ in previous work.[11] In this experiment, the reprecipitation of Zn_2SiO_4 as well as

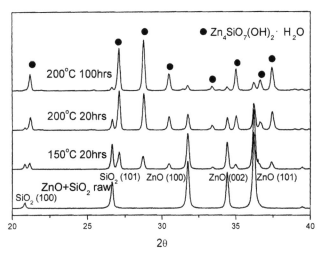

Fig. 1 X-ray diffraction analysis showing stable phases which are present up to 200℃ under autogenous pressure.

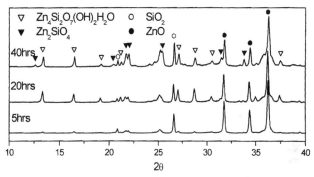

Fig. 2 X-ray diffraction analysis showing the formation of Zn_2SiO_4 phase at 200℃ under autogenous pressure when the molar ratio of $ZnO:SiO_2$ was adjusted to 1.95:1.

$Zn_4Si_2O_7(OH)_2 \cdot H_2O$ phase occurred simultaneously from the initial stage of hydrothermal synthesis at 200℃ as is evident from the XRD data of 5h synthesis in Fig. 2. After synthesis for 20 and 40h, the formation of Zn_2SiO_4 and $Zn_4Si_2O_7(OH)_2 \cdot H_2O$ phase occurred continuously.

As the deficiency of ZnO in the raw materials composition increased, the formation rate of Zn_2SiO_4 increased. This is apparent from the SEM micrographs

(A)　　　　　　　　　　　　　(B)

(C)　　　　　　　　　　　　　(D)

Fig. 3 SEM micrographs showing the formation of prismatic Zn_2SiO_4 particles from different compositions of raw materials. The molar ratio of $ZnO:SiO_2$ in the starting powder was 1.98:1(A and B) and 1.80:1(C and D). Hydrothermal synthesis were performed at 200 ℃ under autogenous pressure for (A) 20h, (B) 40h, (C) 20h, and (D) 40h.

shown in Fig. 3. When the molar ratio of $ZnO:SiO_2$=1.98:1 (Fig. 3 A and B), fine grains of raw materials were residual after hydrothermal synthesis at 200 ℃ for 40h. On the contrary, in case of $ZnO:SiO_2$=1.80:1 (Fig. 3 C and D), residual raw materials could not be observed after 40h under identical hydrothermal conditions. This means that the formation of the Zn_2SiO_4 phase proceeds more easily when there is more deficiency of ZnO in the $ZnO-SiO_2-H_2O$ system under these hydrothermal conditions.

The hydrothermal synthesis of various ZnO deficient compositions at 200 ℃ for 100h showed the formation of Zn_2SiO_4 as the stable phase. In this case, $MnCl_2 \cdot 4H_2O$ was added as an activator. The $Zn_4Si_2O_7(OH)_2 \cdot H_2O$ phase appeared

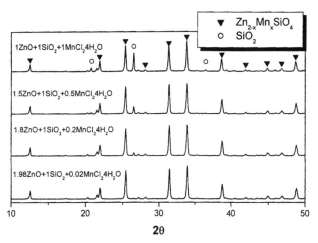

Fig. 4 X-ray diffraction analysis showing the formation of $Zn_2SiO_4:Mn^{2+}$ phase at 200℃ under autogenous pressure. The molar ratio of $ZnO:SiO_2$ in the starting powder was varied from 1.98:1 to 1:1. The amount of $MnCl_2\cdot4H_2O$ added was adjusted to the amount of deficient ZnO deviated from stoichiometric composition.

only up to 40h of synthesis time. A small amount of unreacted SiO_2 phase was also detected by XRD analysis as shown in Fig. 4.

The morphology of Mn doped Zn_2SiO_4 powders was hexagonal prismatic, which was consistent with the shape of $Zn_2SiO_4:Mn^{2+}$ powders synthesized at high temperature-pressure conditions.[11] The SEM micrographs of $Zn_2SiO_4:Mn^{2+}$ particles are shown in Fig. 5. Small particles of SiO_2 phase could be seen along with hexagonal prismatic $Zn_2SiO_4:Mn^{2+}$ particles. The amount of residual SiO_2 particle increased as the mole fraction of SiO_2 in the starting materials increased.

In the hydrothermal synthesis of $Zn_2SiO_4:Mn^{2+}$ phosphor, the actual concentration of Mn doped into the Zn_2SiO_4 host lattice is not consistent with that of the initial Mn addition. ICP analysis for the actual Mn concentrations in $Zn_{2-x}Mn_xSiO_4$ showed that about 10% of the initial Mn content were doped in Zn_2SiO_4 lattice. The efficiency of doping decreased slightly when the initial amount of Mn was doubled. However, it can be said that the absolute amount of Mn doped in the Zn_2SiO_4 lattice increased by increasing the Mn content in the starting composition. These results are shown in Fig. 6 for the composition ranges of $ZnO/SiO2=1.88$ to 1.94.

The PL emission spectra of hydrothermally synthesized $Zn_{2-x}Mn_xSiO_4$ were measured using 254nm-excitation wavelength. The brightness with respect to composition is shown in Fig. 7. The highest brightness of green color was obtained at x=0.02, which is lower than that of $Zn_{2-x}Mn_xSiO_4$ prepared by a solid-

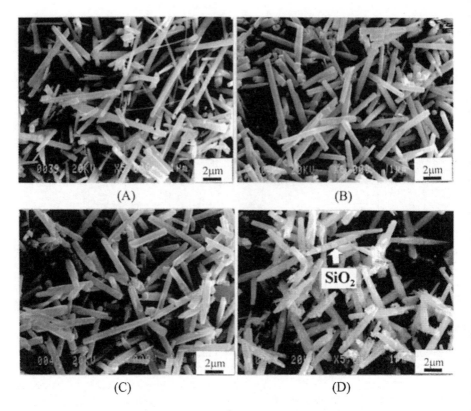

(A)   (B)

(C)   (D)

Fig. 5 Hexagonal prismatic $Zn_2SiO_4:Mn^{2+}$ particles synthesized at 200 ℃ for 100h under autogenous pressure. The molar ratio of ZnO : SiO_2 : $MnCl_2 \cdot 4H_2O$ in the starting raw materials were (A) 1.98:1:0.02, (B) 1.8:1:0.2, (C) 1.5:1:0.5, and (D) 1:1:1.

state reaction.[3] This is due to the fact that the residual amount of SiO_2 particles increases with increase in x value, as can be seen in Fig. 5. However, phosphors of this study showed a brightness of ~55% compared to a commercial one. Thus, optimization of composition as well as synthesis condition should be followed.

CONCLUSIONS

$Zn_2SiO_4:Mn^{2+}$ phosphors were hydrothermally synthesized at 200 ℃ under autogenous pressure. The formation rate of Zn_2SiO_4 phase was affected by ZnO deficiency in the starting composition. The doping efficiency of Mn was found to be ~10% regardless of the amount of Mn addition in the composition range with ZnO/SiO_2=1.88 to 1.94.

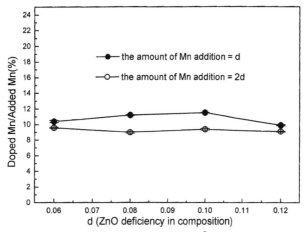

Fig. 6 The doping ratio of Mn in $Zn_2SiO_4:Mn^{2+}$ hydrothermally synthesized at 200℃ under autogenous pressure for 100h. The amounts of Mn addition were d and 2d, where d indicates deficiency of ZnO in the composition of raw materials.

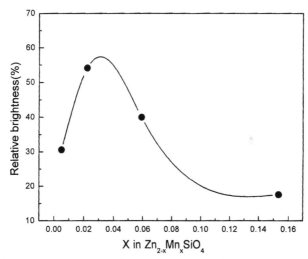

Fig. 7 Brightness changes with respect to Mn doping level of $Zn_{2-x}Mn_xSiO_4$ phosphors hydrothermally synthesized at 200℃ under autogenous pressure for 100h.

PL spectra were measured and brightness changes with respect to composition were investigated. The maximum brightness was observed when the actual Mn content, x, approached 0.02. This phenomenon is attributed to the increase in residual SiO_2 particles at higher x value.

REFERENCES

[1] C.R. Ronda, "Recent Achievements in Research on Phosphors for Lamps and Displays," *Journal of Luminescence*, **72-74** 49-54 (1999).

[2] A.L.N. Stevels and A.T. Vink, "Fine Structure in the Low Temperature Luminescence of Zn_2SiO_4:Mn and $Mg_4Ta_2O_9$:Mn," *Journal of Luminescence*, **8** 443-451 (1974).

[3] A. Morell and N. El Khiati, "Green Phosphors for Large Plasma TV Screens," *Journal of the Electrochemical Society*, **140**[7] 2019-2022 (1993).

[4] C. Barthou, J. Benoit, and P. Benalloul, "Mn^{2+} Concentration Effect on the Properties of Zn_2SiO_4:Mn phosphors," *Journal of the Electrochemical Society*, **141**[2] 524-528 (1994).

[5] R.J.R.S.B. Bhalla, E.W. White, and R. Roy, "Quantitative Electron Microprobe Analysis of Commercial Microcrystalline Phosphor Powders," *Journal of Luminescence*, **6** 116-124 (1973).

[6] S. Hirano, "Hydrothermal Processing of Ceramics," *Ceramic Bulletin*, **66**[9] 1342-1344 (1987).

[7] W.J. Dawson, "Hydrothermal Synthesis of Advanced Ceramic Powders," *Ceramic Bulletin*, **67**[10] 1673-1678 (1988).

[8] T. Toshyuki, "Method of Preparation of Zincsilicate Phosphor Powders," Japanese Pat. No. 63 196683, Aug. 15, 1988.

[9] T. Toshyuki, "Method of Preparation of Zincsilicate Phosphor Powders," Japanese Pat. No. 01 272689, Oct. 31, 1989.

[10] Q.H. Li, S. Komarneni, and R. Roy, "Control of Morphology of Zn_2SiO_4 by Hydrothermal Preparation," *Journal of Materials Science*, **30** 2358-2363 (1995).

[11] C. Yoon and S. Kang, "The Synthesis of $Zn_{2-x}Mn_xSiO_4$ Phosphors Using a Hydrothermal Technique," *Journal of Materials Research*, **16**[4] 1210-1216 (2001).

FeSi$_x$O$_y$ FILMS PREPARED BY CO-SPUTTERING

Takashi Ehara, Masafumi Saito, Shuichi Naito and Yoshihiro Kokubun
School of Science and Engineering, Ishinomaki Senshu University
Shinmito 1, Minamisakai, Ishinomaki, Miyagi 986-8580, Japan

ABSTRACT

In this paper, preparation and properties of Iron-oxysilicide (FeSi$_x$O$_y$) by reactive co-sputtering method will be described. Reactive sputtering by oxygen gas using Fe pellets placed on Si target produces Iron-oxysilicide thin films. The films showed an optical bandgap of 1.4 to 2.8 eV that is larger than that in β-FeSi$_2$. In the absorption spectra, localized states were observed. The optical properties observed were due to Fe atoms included in the films. In addition, the films displayed low enough conductivity to use as an insulating material. The results observed in the present work indicate that the Iron-oxysilicide films are preferable to use in β-FeSi$_2$ based devices as an insulating layer.

INTRODUCTION

Recently, beta-irondisilicide (β-FeSi$_2$) thin films have gathered much interest as a novel semiconductor.[1-3] The β-FeSi$_2$ has an optical bandgap of 0.8 – 0.9 eV that is preferable for optical fiber communication systems.[1,4-6] In the use of the β-FeSi$_2$ for devices, adequate insulating material is required. However, use of conventional insulating materials, such as SiO$_2$ or SiN$_x$, may induce diffusion of

---

Fe from β-FeSi$_2$ to the insulating layer. An insulating material whose characteristics are not affected by diffusion of Fe will be requested. We think the materials that include a sufficient amount of Fe in the films may not be affected by diffusion of Fe atoms. Previously, present authors have reported thermal oxidation of β-FeSi$_2$ thin films.[5] The oxidized layer showed electronic properties as an insulating layer. However, the layer formed by thermal oxidation did not show enough uniformity to use as an insulating layer for devices.

In the present work, we have prepared oxidized insulating layer for β-FeSi$_2$. Preparation of the films has been carried out by reactive sputtering using the same sputtering target, consisting of Fe pellets and a Si target, as in the preparation of β-FeSi$_2$. Optical and electrical properties of the films will be described.

EXPERIMENTAL DETAILS

The FeSi$_x$O$_y$ thin film samples were prepared by the reactive rf co-sputtering method. Adequate pieces of Fe pellets with 10 mm in diameter were placed on a Si target (100 mm in diameter) and co-sputtered with O$_2$ gas at a 6 sccm feed rate. Deposition was carried out at sputtering pressure of 1.33 Pa and rf power of 100 W (13.56 MHz) for 0.5 hour using an HSR551 sputtering machine (Shimadzu). We have measured the properties of as prepared and thermally annealed films. It is because we expect the β-FeSi$_2$ base device fabrication that includes thermal annealing of insulating layer. The annealing was carried out at 800°C for 0.5 hour in nitrogen atmosphere. Thin films at a thickness of at least 0.1 μm were deposited on both SiO$_2$ and Si substrates.

Infrared (IR) absorption spectra were measured to study the structural properties of the films. IR spectra were obtained using a Fourier transform-infrared (FT-IR) spectrometer (Shimadzu 8100A). Structure and electronic properties of the films were checked by X-ray diffraction and measurement of conductivity, respectively. X-ray diffraction was measured using CuK$_\alpha$ radiation. Conductivity of the films were examined using Al interdigital electrode that has 0.2 μm of gap and 195 mm of length.

RESULTS AND DISCUSSION

Before the preparation of the oxide films, we have prepared oxygen free films using Ar as a sputtering gas to determine the adequate amount of Fe pellets. Due to the investigation, deposition using eight pellets is adequate for preparation of $FeSi_2$. As the oxide layer studied in the present work will be used for β-$FeSi_2$ base device, the films that are prepared around this condition should be studied.

In the figure 1, optical absorption spectra of the films prepared using 6 to 10 pellets are shown. As the films were amorphous, the absorption spectra were plotted by Tauc plot. The bandgap determined by optical absorption shown are from 2.2 to 2.5 eV. The absorption spectra showed increase of $(\alpha h\nu)^{1/2}$ at photon energy of 2 eV. Although it is not shown, the result does not depend on the thickness of the films. The result indicates that the absorption at around 2 eV is due to the localized states in the films. In the figure 2, absorption spectra of the thermally annealed samples are shown. According to the measurement of X-ray diffraction, the films were still amorphous after the annealing. Thus, the absorption spectra were described by a Tauc plot. The absorption spectra changed drastically after the thermal annealing. The bandgap after the annealing is from 1.5 to 1.7 eV. In addition, gradients of the Tauc plot have also been decreased. This result indicates a change in band edge by annealing. In the figure 3, the optical bandgap determined from figure 1 and 2 are shown. In either before or after the annealing, optical bandgap does not show strong dependency on chemical composition of the films.

In the absorption spectra, existence of localized states has been observed. We have investigated the details of the absorption at photon energies of 1.1 to 2.0 eV. In figure 4, absorption spectra of the sample prepared using 8 pellets are shown before and after the annealing. In the figure, vertical axis is transmittance. In the figure, change of the localized states by annealing is observed as a change in the absorption. After the annealing, two strong and sharp peaks at 1.72 and 1.80 eV were observed. The result means there are two major localized states in the gap. The localized states might affect the electronic properties of the films. In figure 5, the absorption spectra of the annealed sample prepared by 6, 8 and 10 Fe pellets from 1.1 to 2.2 eV are shown. As shown in the figure, the transmission of the localized states depends on the amount of Fe pellets. The two peaks observed in the sample prepared using 6 or 8 pellets become one peak in the sample using 10 pellets. The result indicates that the localized states are induced by the existence of Fe in the films.

Structural properties have been studied using IR absorption spectra. In figure 6, IR spectra between 700 and 1400 cm^{-1} of the sample prepared using 6, 8 and 10 Fe pellets before annealing are shown. In the IR spectra, only two peaks have been observed in all samples. One is a weak Si-O-Si bending at 800 cm^{-1} and the other is Si-O-Si stretching at 1080 cm^{-1}. The IR peaks observed in the present work is essentially consistent with that in SiO_x prepared by sputtering method.[6] The sharp peak showed clear dependency on chemical composition. The peak width increased with increase in Fe in the films. This is because the Fe atoms in the films affect the bonding properties of Si-O-Si.

Finally, we have tried to measure the conductivity of the films. However, the conductivity of the films was too low to measure. We also have fabricated MOS devices consisting of β-$FeSi_2$/$FeSi_xO_y$/Al. Thickness of the β-$FeSi_2$ and $FeSi_xO_y$ layers are 1.2 μm and 0.1 μm, respectively. The $FeSi_xO_y$ layers have not been annealed. The device worked as a capacitor and the result means the $FeSi_xO_y$ layer worked as an insulating layer. The authors think the device can work as a tunneling diode if the thickness of the insulating layer is thinner.

CONCLUSION

The $FeSi_xO_y$ films were prepared by reactive co-sputtering. The optical and electrical properties observed are adequate as an insulating layer for β-$FeSi_2$ base devices.

AKNOWLEDGEMENT

The authors express special thanks to Prof. Shinji Nakagomi for his helpful advises and discussion.

REFERENCES

[1]M. C. Bost and J. E. Mahan, "Optical properties of semiconducting $FeSi_2$ films," *Journal of Applied Physics* **58** 2696-2703 (1985)

[2]K. Okajima, H. Yamatsugu, C. Wen, M. Sudoh and K. Yamada, "Spectral sensitivity enhancement by thin film of β-$FeSi_2$-Si composite prepared by RF-sputtering deposition," *Thin Solid Films* **381** 267-275 (2001).

[3]D. Leong, M. Harry, K. J. Reeson and K. P. Homewood, "A silicon/iron-disilicide light emitting diode operating at a wavelength of 1.5 _ m" *Nature* **387** 686-688 (1997).

[4]C. A. Dimitriadis, J. H. Werner, S. Logothetidis M. Stutzmann, J. Weber and R. Nesper, "Electronic properties of semiconducting $FeSi_2$ films." *Journal of Applied Physics,* **68** 1726-1734 (1990).

[5]N. E. Christensen, "Electronic structure of β-$FeSi_2$," *Physical Review B"* **42** 7148-7153 (1990).

[6]D. Leong, M. Harry, K. J. Reeson and K. P. Homewood, "On the origin of the 1.5 micron luminescence in ion beam synthesized beta-$FeSi_2$," *Applied Physics Letters* **68** 1649-1650 (1996).

[7]T. Ehara, Y. Sasaki, K. Saito, S. Nakagomi and Y. Kokubun, "β-$FeSi_2$,-base MIS diode fabricated by sputtering method," *Applied Surface Science*, in press

[8]T. Ehara and S. Machida, "The Effect of Nitrogen Doping on the Structure of Cluster or Microcrystalline Silicon Embedded in thin SiO_2 films" *Thin Solid Films,* **346** 275-279 (1999).

FIGURE CAPTIONS

Figure 1: Absorption spectra of as prepared $FeSi_xO_y$ films prepared using 6, 7, 8, 9 and 10 Fe pellets. The spectra are shown using a Tauc plot.

Figure 2: Absorption spectra of $FeSi_xO_y$ films prepared using 6, 7, 8, 9 and 10 Fe pellets after the thermal annealing at 800°C.

Figure 3: Optical bandgap of $FeSi_xO_y$ films prepared using 6, 7, 8, 9 and 10 Fe pellets before and after the annealing.

Figure 4: Transmission spectra of $FeSi_xO_y$ films prepared using 8 Fe pellets, before and after the annealing.

Figure 5: Transmission spectra of FeSi$_x$O$_y$ films prepared using 6, 8 and 10 Fe pellets after the annealing.

Figure 6: Infrared absorption spectra of FeSi$_x$O$_y$ films prepared using 6, 8 and 10 Fe pellets after the annealing.

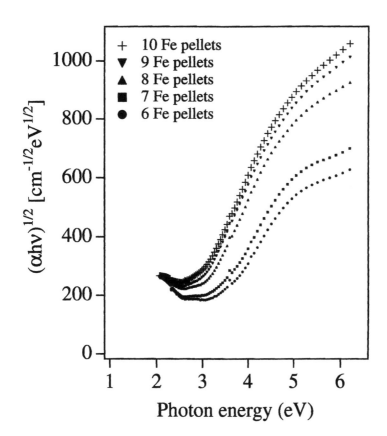

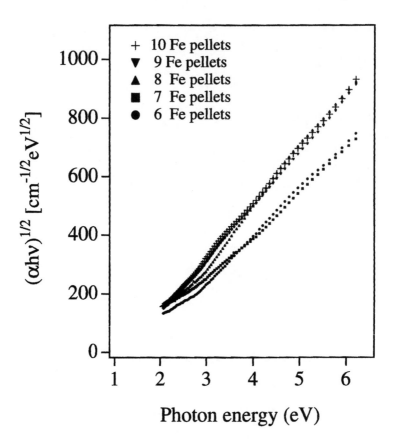

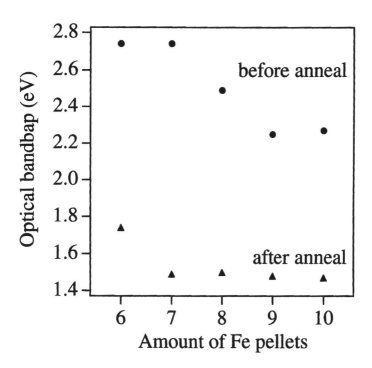

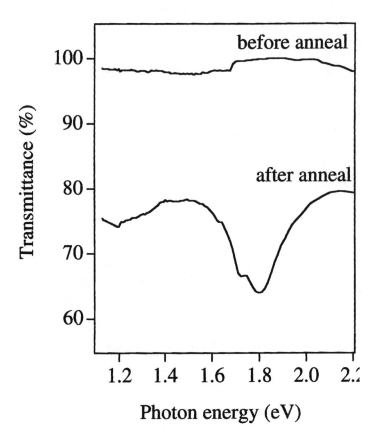

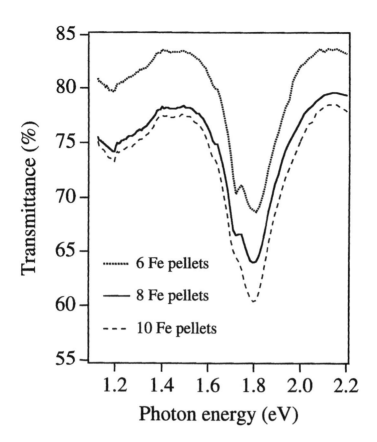

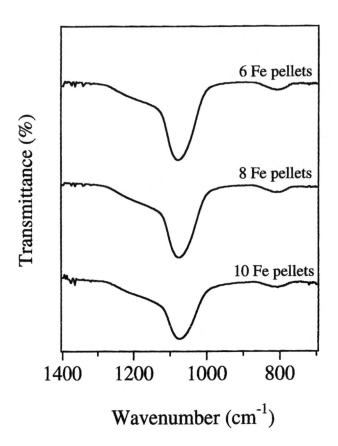

Structure-property relationships in As-S-Se glasses for waveguide applications probed by near-infrared Raman spectroscopy

Clara Rivero[1], Alfons Schulte, Kathleen Richardson
School of Optics – CREOL and Department of Physics,
University of Central Florida, Orlando, FL 32816-2385, USA

ABSTRACT

Thin film devices based on chalcogenide glasses (ChG) are attractive for integrated optics applications due to their good infrared transmission and high nonlinear Kerr effects. We structurally characterize variations in physical properties in As-S-Se device structures depending on material processing and in-use laser conditions by near-infrared micro and waveguide Raman spectroscopy. At S/Se=1 molar ratio and with decreasing As content (coinciding with the compositional range of high nonlinear optical coefficients) new vibrational bands appearing around 255 cm^{-1} and 440 - 480 cm^{-1} are attributed to Se-Se and S-S homopolar bonds and formation of a layer-like glass network. Near-infrared Raman spectroscopy of photoinduced and annealed structurres allows to identify specific bonding changes which accompany the aging process.

INTRODUCTION

Materials suitable for use in high speed optical communication applications require all-optical processing and switching capabilities which necessitate they be compatible with current system configurations, possess ultrafast broadband response time, as well as low linear and nonlinear loss. Additionally, the material must be amenable to the small sizes and stability requirements needed for future "on chip" device configurations. Chalcogenide glasses (ChGs) have shown promise in that they exhibit properties compatible with the above requirements at 1.3 and 1.55 µm wavelengths [1]. Recent experiments on As-S, As-S-Se and related compounds have demonstrated flexibility in forming fiber and film devices such as narrow-band spectral Bragg filters [2], lens-arrays as well as couplers and self-written planar waveguides made from single and multilayer film structures [3]. Efforts to optimize film properties and device performance have focussed on identifying the chemical and structural origin of the linear and nonlinear response in terms of the material processing conditions used in creating the optical element [4]. ChG's are photosensitive when exposed to bandgap energy (Eg ~2.35 eV for As$_2$S$_3$) [5,6]. Use of these photosensitive effects (photodarkening and photoexpansion) in

---

[1] Presented at the A2- Optoelectronic Materials and Technology in the Information Age program, 103rd Annual Meeting of the American Ceramics Society.

ChG's, allows the creation of bulk waveguide structures [7], or the patterning of photoinduced relief gratings and guided wave structures in ChG films [8].

Raman spectroscopy employing integrated optical techniques can be extremely powerful in the microstructural analysis of thin film devices due to the combination of good molecular specificity and high sensitivity. The material of interest is cast into a slab waveguide thereby significantly increasing both the scattering volume and the electrical field intensity within the film. In spite of its sensitivity waveguide Raman spectroscopy (WRS) using guided mode excitation [9] has not been applied to the structural characterization of chalcogenide glasses, most likely due to their high index (~ 2.45) and lack of suitable prism couplers. In addition, while ChGs materials are technologically important in integrated optics applications at infrared telecommunication wavelengths, they can undergo photostructural modifications when Raman scattering is excited with wavelengths in the visible. Here, cleaved silicon substrates are employed for high efficiency end coupling of a near-infrared laser beam into a ChG layer structure. The NIR Raman spectra of the waveguide provide molecular structural information on material optical properties representative of those observed during device operation.

EXPERIMENTAL

Single and multilayer waveguides were deposited via thermal evaporation of As_2S_3 bulk glasses on (100) silicon substrates as previously described [3]. Waveguides examined were nominally $1.75 - 2$ µm thick and unless otherwise stated were measured in their as-deposited (unannealed) form. An 840 nm output beam from a continuous mode Ti:sapphire laser with 25 mW power was end coupled into a 2 µm As_2S_3 film via of a 20 X microscope objective. The energy entering the guide was less than the above as the spot size of the incident beam was larger than the cross-section of the film. Propagation was verified by imaging the mode at the waveguide output. The 90° Raman scattered light emanating from the top surface of the film under examination was collected with a lens and spectrally analyzed with a back-illuminated, thinned CCD detector mounted on the exit port of a single grating spectrograph. The Rayleigh line was suppressed with a CdTe bandgap filter.

DATA ANALYSIS AND RESULTS

Molecular subunits in As_2S_3 bulk glasses and films

The Raman spectra obtained from bulk As_2S_3, an optical fiber, and an as deposited film are shown in Fig. 1. Scattering was excited at sub bandgap energy (840 nm) using 90° or waveguide Raman geometry. The dominant feature in the As_2S_3 spectra is a band at 345 cm^{-1} attributed to an antisymmetric As-(S)-As stretching vibration in the $As(S)_{3/2}$ pyramids. According to an analysis by Lucovsky and Martin [10] the normal modes of the bulk glasses (e. g. clusters of $As(S)_{3/2}$ molecules with weak intermolecular coupling) are obtained by treating the molecular $As(S)_{3/2}$ and bridging chain modes (As-(S)-As) independently. A detailed comparison of the sub-components of the broad bands with spectral signatures of possible structural units predicted by molecular modeling is in progress.

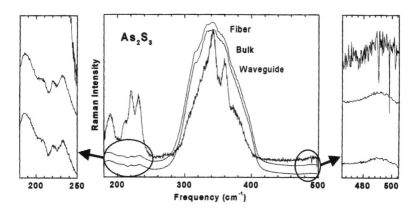

Fig. 1. Raman spectra of As_2S_3 bulk glass, fiber, and film (waveguide) measured with 1.5 cm^{-1} spectral resolution. The excitation wavelength is 840 nm.

Though films have less than 1.5 μm thickness high signal-to-noise is achieved by guided mode excitation [11]. The distinction observed between bulk and fiber spectra and that of the film structures emphasizes the resulting structural differences associated with the processing conditions. It can be noted that, while key structural features such as the band at 345 cm^{-1} remain in all spectra, the substructure within the film spectra as compared to the broad features of the bulk and fiber spectra result from different arrangements of the constituent atoms. These sharp, molecular-signature structures were confirmed not to be due to crystallinity within the film, but most likely result from the formation of as-deposited As_4S_4 units [12]. A strong polarization dependence of the Raman scattered light is observed when the electric field vector of the exciting beam is in the plane of the waveguide. This allows to discriminate the As_4S_4 units from the structural components that are similar to the ones in the bulk glasses but present in much lower concentration (< 20%). Such differences in local configuration would be expected to affect both the linear and nonlinear properties of the resulting waveguide structure. Subtle variations in structure induced by the forming (film drawing) process can be seen in the weak shoulders of the fiber spectra on the high wavenumber side of the 345 cm^{-1} band.

Compositional effects on molecular structure and nonlinear optical properties

The introduction of selenium to the As-S glass system increases the nonresonant non-linear refractive index, n_2, up to 400 times the value for fused silica [4]. The largest increase in n_2 was observed in a glass whose composition is $As_{24}S_{38}Se_{38}$ [4]. Fig. 2 depicts the near-infrared Raman spectra of As-S-Se glasses in the compositional range where high nonlinear optical coefficients are found. The binary sulfide and selenide compounds, shown for comparison, exhibit strong bands near 345 cm^{-1} and 230 cm^{-1}. A commensurate increase and decrease of these bands is observed in $As_{32}S_{34}Se_{34}$. Overall, similar features emerge here in both, bulk glasses and films. While the As_2S_3 spectra (Fig. 1) demonstrate that the film structure is composed of both individual molecular units and polymeric crosslinked units, Fig. 2 shows that these molecular subunits change substantially with stoichiometry. The Raman spectra of the chalcogen-rich films (Fig. 2) display a reduction of the sharp, molecular-signature structures.

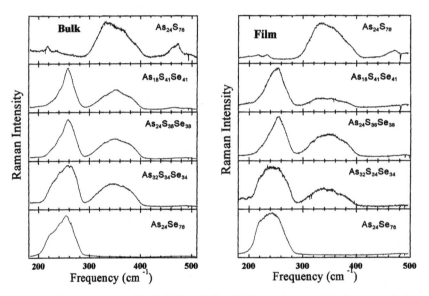

Fig. 2. Raman spectra of bulk (left) and film (right) chalcogen-rich glasses excited at 840 nm.

In the ternary compounds with S/Se = 1 molar ratio and decreasing As content, narrower bands around 255 cm^{-1} and 440 - 480 cm^{-1} form, and are attributed to Se-Se and S-S homopolar bonds. The preferential formation of these covalent bonds is correlated with the appearance of enhanced non-linear optical properties. The progressive decrease of the relative intensities of bands characteristic of the As-S or As-Se structure emphasizes the disappearance of pyramidal sites in the layer-like glass network, while homopolar bonds form chalcogen chains connecting the remaining pyramidal units. The small number of S-S bonds indicated by a weak band near 495 cm^{-1} for equal concentrations of S and Se, suggests that the S stays with the remaining pyramids, and it is the Se which dominates the connecting chain units and the nonlinear refractive index.

Annealing and photoinduced structural changes

The ultimate long-term stability of chalcogenide based optical elements relies on the generation of photoinduced structures which undergo limited structural relaxation with time. ChG's typically exhibit lower glass transition temperatures (Tg) than oxide glasses and hence, can exhibit significant sub-Tg relaxation at, or near, room temperature [13]. This relaxation can result in structural changes that modify the *as-written* glass structure and performance of the optical element. To ascertain the effects of structural relaxation resulting from aging, spectroscopic analyses of As$_2$S$_3$ film structures were performed following annealing and in freshly written (< 1 month) and aged (~ 3 years under ambient conditions) Bragg gratings. In all cases, a depletion in as-deposited/as-written As-As and S-S bonds and modification of As-S bonds, was observed. The extent of such changes depended on the initial, as-formed concentrations of each, and how the film was aged or annealed. Fig. 3 illustrates such changes in waveguide Raman spectra for As$_2$S$_3$ structures prior to and after annealing, and following 514 nm exposure similar to that used in grating writing.

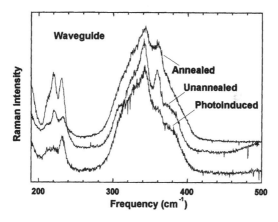

Fig. 3. Variation in waveguide Raman spectra (excited at 840nm) for fresh, annealed and photostructurally modified As$_2$S$_3$ single layers (exposure λ=514.5nm)

The sharp features seen in the as-deposited Raman spectrum are not present in bulk GhG's and correspond to molecular units that can undergo reconfiguration over time. While small amounts of the *as-written* relief structure stayed in the *aged* grating structure, few molecular species seen in as-deposited structures, remained. The effects of photo-induced changes are evident in the Raman spectra as well. By exciting Raman scattering with sub-bandgap energies (λ_{exc}= 840 nm) at low power levels (25 mW) photostructural changes due to the probe beam are avoided. Large changes are seen in the Raman band near 235 cm^{-1} which is associated with As-As bond vibrations. Several mechanisms for photoinduced changes in chalcogenides, including bond redistribution and coordination defects, have been suggested [14]. The impact of such structural changes on expected grating optical performance will be assessed.

CONCLUSIONS

In conclusion, near-infrared waveguide Raman spectroscopy is employed to structurally characterize chalcogenide thin film devices *in-situ* revealing differences in the molecular subunits as a function of the S/Se ratio. The formation of homopolar S-S and Se-Se bonds is observed in the compositional range where high nonlinear optical coefficients are found. Further studies will aim to dynamically probe the structural rearrangement associated with specific film annealing conditions. It is also expected that the evolution of a self-written channel during exposure to strong laser pulses could be characterized in-situ using the WRS technique, as well as photostructural changes, which occur during illumination.

ACKNOWLEDGMENTS

This work was carried out with the support of a number of research, equipment and educational grants, including NSF DMR-9974129, NSF REU grants EEC-9732420 and CHE-9732161, NSF DUE-9850934. We appreciate the use of fiber provided by Ray Hilton, Sr., Amorphous Materials, Inc., Garland TX., and the high purity As$_2$S$_3$ powder synthesized by professor Frumar's group in the Czech. Republic. Also, Clara Rivero acknowledges a fellowship from the Honor's College at the University of Central Florida.

Finally, thanks to Cedric Lopez, Gero Nootz, and all other members of the chalcogenide group at the University of Central Florida/CREOL; in addition to T. Galstian, A. Villeneuve, R. Valee, V. Hamel, and K. Turcotte, our collaborators at Laval University in Canada.

REFERENCES

1 K. A. Richardson, J. M. McKinley, B. Lawrence, S. Joshi, and A. Villeneuve, J. Optical Materials **10**, 155-159 (1998).

2 K. Tanaka, N. Toyosawa, and H. Hisakuni, Opt. Lett. **20**, 1976-1978 (1995).

3 J.-F. Viens, C. Meneghini, A. Villeneuve, T. Galstian, E. J. Knystautas, M. A. Duguay, K. A. Richardson, and T. Cardinal, J. Lightwave Technology **17**, 1184 (1999).

4 T. Cardinal, K. A. Richardson, H. Shim, G. Stegeman, R. Beathy, A. Schulte, C. Meghini, J. F. Viens, and K. Le Foulgoc, A. Villeneuve , J. Non-Cryst. Solids **256 & 257**, 353-360 (1999).

5 T.V. Galstyan, J.-Duguay, K. Richardson, J. of Lightwave Technology, vol. 15, pp. 1343-1347, 1997

6 S. Ramachandran, S.G. Bishop, J.P. Guo, D.J. Brady, IEEE Phot. Tech. Lett., vol. 8, pp. 1041-1043, 1996

7 O.M. Efimov, L.B. Glebov, K.A. Richardson, E. Van Stryland, T. Cardinal, S.H. Park, M. Couzi, J.L. Bruneel, in press, J. Opt. Mater., (2000)

8 A.M. Andriesh, Yu. A. Bykovskii, E.P. Kolomeiko, A.V. Makovkin, V.L. Smirnov, and A.V. Shmal'ko, Sov. J. Quantum Electron., vol. 7, pp. 347-352, 1977; M. Asobe, T. Kanamori, and K. Kubodera, IEEE J. Quant. Electr., vol. 29, pp. 2325-2333, 1993; S. Ramachandran, S.G. Bishop, Appl. Phy. Lett., vol. 74, pp. 13-15, 1999

9 J. F. Rabolt, in *Fourier Transform Raman Spectroscopy*, Eds. D. B. Chase and J. F. Rabolt, (Academic Press, San Diego, 1994), p. 133-157.

10. G. Lucovsky and R. Martin, J. Non-Cryst. Solids **8-10**, 185-190 (1972).

11 A. Schulte, C. Rivero, K. Richardson, K. Turcotte, V. Hamel, A. Villeneuve, T. Galstian, R. Valle, Opt. Comm., (submitted)

12 A. J. Apling, A. J. Leadbetter, and A. C. Wright, J. Non-Cryst. Solids **23**, 369-384 (1977).

13 K.A. Cerqua-Richardson, Ph.D; Thesis, NYS College of Ceramics, Alfred University (1992)

14 M. Frumar, M. Vlcek, Z. Cernosek, Z. Polak, T. Wagner, J. Non-Cryst. Solids **213&214**, 215-224 (1997).

STUDY OF STRUCTURAL CHANGES IN GLASSY As$_2$Se$_3$ BY EXAFS UNDER *IN-SITU* LASER IRRADIATION

Gang Chen and Himanshu Jain
Department of Materials Science & Engineering
Lehigh University
5 East Packer Avenue
Bethlehem, PA 18015-1539

Syed Khalid
Brookhaven National Lab
NSLS Bldg. 725D,
PO Box 5000
Upton, NY 11973-5000

Jun Li and David A. Drabold
Department of Physics and Astronomy
Ohio University
Athens, OH 45701-2979

Stephen R. Elliott
Department of Chemistry
University of Cambridge
Cambridge, England

ABSTRACT

The properties and structure of chalcogenide glasses are known to be sensitive to the light of bandgap energy. We have used *in-situ* extended x-ray absorption fine- structure (EXAFS) analysis to study the changes in local atomic structure, which are induced by laser (wavelength= 690 nm) irradiation of glassy As$_2$Se$_3$. The results indicate the atomic configuration of both the temporary and permanent light-induced changes in the chalcogenide glass.

INTRODUCTION

Semiconducting chalcogenide glasses (e.g., Se, sulfides or selenides of Ge, Sb or As, etc.) have been studied for many years because of the variety of metastable changes that they exhibit when illuminated by bandgap or sub-bandgap light.[1,2] Under irradiation, pairs of free electrons and holes are created, which become trapped or localized in amorphous semiconductor because of the strong electron-phonon interaction. This process leads to changes in local structure of materials, and hence also in the physical properties causing, for example, photodarkening, photocrystallization, photoexpansion, etc.[3-6] Recently reported opto-mechanical effect that could be applied to nanotechnology indicates the importance of photostructural changes [7].

Extended x-ray absorption fine structure (EXAFS) is one of the very few experimental techniques that can provide atomic scale information regarding the photo-induced structure changes in the semiconducting materials. It is sensitive to local atomic structure and hence can be a powerful tool for probing the photostructural changes in chalcogenides [8,9]. Most previous EXAFS experiments for studying photostructural changes were conducted *ex situ*, where the structure of glass was determined in the as prepared stage and then after the sample was irradiated separately [10-12]. So the information about temporary structural changes "during irradiation" was not detected. Recently Kolobov et. al reported *in situ* EXAFS experiments on a-Se film at 30K. They observed light-induced increase of average coordination number (about 5%) during Xe lamp illumination [9]. They attribute this change to the formation of dynamic interchain Se-Se bonds via interaction of excited lone-pair electrons. In this paper we report the temporary as well as irreversible or permanent light-induced changes in glassy As_2Se_3 as observed by *in situ* Se and As K-edge EXAFS.

EXPERIMENT

Fine glassy As_2Se_3 powder* (99.999% purity on metal basis, 325 mesh) was used for EXAFS spectra, which were measured at National Synchrotron Light Source (NSLS) on the beamline X18B with a Si(111) channel cut double-crystal monochromator. The piezo-driven picomotors detuning and deglitching devices in monochromator significantly improved the stability of incident x-ray. Harmonic rejection was accomplished by detuning to reduce the x-ray to 50% of its maximum level. Data were taken in both transmission and fluorescence mode. To eliminate sample inhomogeneity, which is the largest source of noise in transmission mode, we weighed certain amount powder, pressed and sealed it between two Scotch tapes. Then several such layers were put together to optimize the signal-to-noise ratio. The effective sample thickness was estimated to be 20-30 μm. Passivated Implanted Planar Silicon (PIPS) detector was used for measuring fluorescence signal. An aluminized Mylar sheet was placed on the PIPS detector to prevent noisy light from influencing the detector.

The experiment setup is shown schematically in Fig. 1: The sample was mounted on a copper holder in a cryostat cylinder equipped with kapton windows for passing x-ray and laser light. A 690nm diode laser with 30mW maximum output power was used for in situ illumination, and the light intensity on the sample is estimated to be 100-150mW/cm^2. The laser diode driver was extended outside of the beamline hutch so that we could operate laser with a remote control. Standard As foil was placed before the reference ionic chamber to calibrate As K-edge of As_2Se_3 for each scan.

---

* Alfa Aesar, A Johnson Matthey Company, Ward Hill, MA, USA

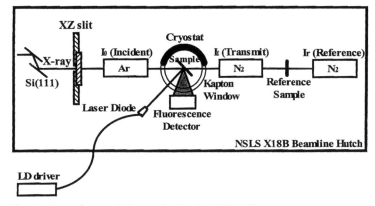

Fig. 1. Experimental Setup for In situ EXAFS

The EXAFS spectra were obtained for the As K-edge (11867 eV) and Se K-edge (12658 eV) in the same scan in transmission and fluorescence at room and 40K temperatures. The procedure for *in situ* laser irradiation was as follows: first take several consecutive spectra in dark, then turn on the laser and irradiate the sample for 4 hours recording the spectra at regular intervals. Then, switch off the laser and record the spectra several times without the laser irradiation.

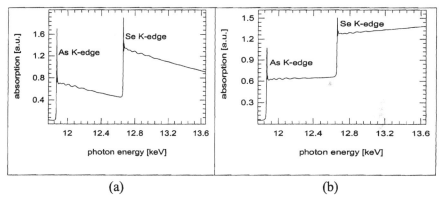

(a)   (b)

Fig. 2. As and Se K-edge transmission (a) and fluorescence (b) spectra for glassy As_2Se_3 at room temperature. These two spectra were taken at the same scan.

Fig.2 shows raw EXAFS spectra at room temperature for both As and Se K-edge in transmission and fluorescence. Good sample homogeneity and stable high-brilliance synchrotron light source as well as high performance ionic chamber and PIPS detector make the precision of *in situ* EXAFS ±0.5% for a given sample under different light exposure conditions. The data show that X-ray exposure alone does not produce any detectable photostructural changes.

DATA ANALYSIS AND RESULTS
The K-edge of the normalized EXAFS can be described by [13]

$$\chi(k) = -\sum_i A_i(k) \sin[2kR_i + \varphi_i(k)] \tag{1}$$

where the summation extends over i coordination shells at average distance R_i from the absorption atom, k is the photoelectron wave vector, and $\varphi_i(k)$ is the total phase shift due to contributions from both the absorbing and the backscattering atoms. The amplitude function $A_i(k)$ is given by

$$A_i(k) = \frac{N_i}{kR_i^2} F_i(k) \exp(-2\sigma_i^2) \exp(-2R_i / \lambda) \tag{2}$$

where N_i is the average number of scattering atoms, $F_i(k)$ is the backscattering amplitude characteristic of a particular type of scattering atom, σ_i^2 is the Debye-Waller factor or mean square relative displacement (MSRD) that accounts for thermal vibrations and static disorder, and λ_i is the mean-free path of the photoelectron.

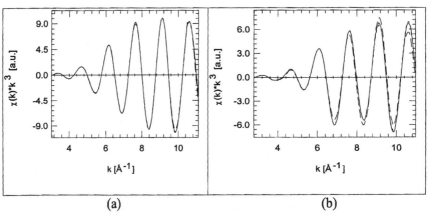

(a)   (b)

Fig. 3. Plots of EXAFS As K-edge (a) and Se K-edge (b) oscillation, $\chi(k)*k^3$, plotted as a function of photelectron vector k before laser illumination (solid line), during illumination (dashed line) and after illumination (dotted line). Note that the 'before' and 'after' illumination curves are almost overlapping each other.

Expression (1) and (2) are used by WINXAS program for data refinement and simulation. Backscattering amplitude $F_i(k)$, phase shift $\varphi_i(k)$ and λ_i are taken from FEFF program output file, which is derived from a standard *ab initio* multiple scattering calculation [14]. The input file based on the As_2Se_3 crystal structure [15] for FEFF is created by ATOMS software.

$\chi(k)$ EXAFS oscillations for As and Se K-edge before, during and after laser illumination were extracted from raw spectra as shown in Fig. 3. EXAFS oscillations $k^3\chi(k)$ were Fourier transformed using the region from k=2.7 to 11.7 Å^{-1} for the As K-edge, and from k=3 to 12.5 Å^{-1} for Se K-edge. The transformed spectra are shown in Fig. 4, and the structural parameters obtained by FEFF fitting are shown in table I.

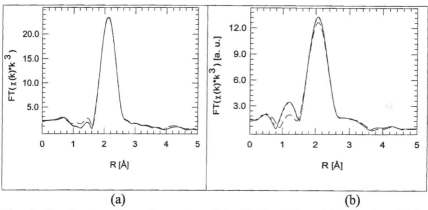

(a)   (b)

Fig. 4. Fourier transformed spectra of As K-edge (a) and Se K-edge (b) before (solid line), during (dashed line) and after laser illumination (dotted line).

Table I. Results of the fits to As and Se EXAFS before laser illumination (BI), during illumination (DI) and after illumination (AI)

	As(BI)	As(DI)	As(AI)	Se(BI)	Se(DI)	Se(AI)
R(X) ±0.001	2.418	2.418	2.418	2.405	2.406	2.405
Normalized CN ±0.005	1	0.999	0.998	1	0.983	0.996
Normalized MSRD ±0.005	1	0.996	1.002	1	1.029	1.009

The experiments were repeated on several samples, and the results of the observed changes are highly reproducible. Concerning the error bar in our *in situ*

EXAFS experiments, note that the coordination number error is mainly caused by position of the relative to x-ray beam. Since we did not move the sample during the whole experiment and the x-ray beam position is fixed, the error should be very small as confirmed by reproducibility of results within ±0.5%.

DISCUSSION

From Fig. 4 we note that during bandgap laser illumination, there is no detectable change around As atoms, while the first shell peak around Se atoms decreases reversibly. By comparing our Fourier transformed spectrum for the Se K-edge with that of a-Se film obtained by another group [9], we find a difference: in the present case, the first shell peak of the spectrum decreases reversibly, whereas for a-Se, it increases during photoexcitation. That means different mechanism may be operating for these two cases.

According to the fitting results in Table I, there is no change in the average nearest neighbor distance for As and Se atoms. We also find that during illumination, the average coordination number (CN) of Se atoms decreases (1.7%) and the Debye–Waller factor (σ^2) or so-called mean square relative displacement (MSRD) of Se increases (2.9%) whereas no such change occurs around As atoms. Here we would like to note the correlation between the fitting parameters CN and MSRD. To isolate the correlation for the "during illumination" data of Se K-edge, we first fixed the CN, then the fitting gave a 3.6% increase in MSRD. Next, we fixed MSRD, and obtained a decrease of 3.4% in the CN. So we conclude that average CN of Se atoms definitely decreases during laser illumination. By applying the same fitting to As K-edge, we find no change in CN and MSRD from laser illumination. Note that the penetration depth of 690 nm laser light on the glassy As_2Se_3 is at most half of the sample thickness, so the changes should be doubled if we consider the entire sample region scanned by X-rays. Thus during laser illumination the average CN change for Se should be about 4%, and the MSRD change should be 6%. Considering that there is no detectable MSRD change for As atoms, and if we assume there is also no change of MSRD for Se atoms, the CN of Se atoms should decrease by 7%. The decrease in Se CN means these atoms change from high to low coordination during illumination, causing bond-breaking around Se atoms. However there is no change of coordination around As atoms during illumination. So we conclude that the bond breaking happens only in Se-Se bonds. After switching off the laser, most of the broken bond reformed Se-Se bonds, and light-induced structural disorder (MSRD) around Se atoms partly recovered.

In ideal case, the coordination number for As and Se should be 3 and 2 respectively. However, in chalcogenides glasses, the "wrong coordination" always exists. In a-Se film, it is reported that about 20% Se atoms are 3-fold coordinated [9], and first-principles molecular-dynamics calculation shows [16] that in g-As_2Se_3 19.4% and 20.2% of Se atoms are in 3-fold and 1-fold coordination, respectively. However only 1.2% and 2.3% of As atoms are in 2-fold and 4-fold coordination, respectively. Our results indicate that bond-breaking of Se-Se is a major photostructural change in glassy As_2Se_3 caused by bandgap light.

CONCLUSION

Reversible photostructural changes have been observed by in situ EXAFS. Under bandgap laser illumination, no change occurs in the first shell average distance around Se and As atoms. The major light-induced change is in the breaking of Se-Se bonds, which causes a decrease of about 4% in the Se coordination number. The broken Se bonds mostly reform but a part of the structure disorder induced by light persists after laser is switched off.

ACKNOWLEDGEMENT

The authors gratefully thank the National Science Foundation for supporting this work through the Focused Research Group grant DMR-0074624.

REFERENCES
[1]K. Shimakawa, A. Kolobov, and S. R. Elliott, "Photo-induced effects and metastability in amorphous semiconductors and insulators," Adv. Phys. **44**, 475 (1995).
[2]K. Tanaka, "Photoinduced structural changes in amorphous semiconductor," Semiconductors **32**, 861 (1998).
[3]A. Ganjoo, K. Shimakawa, H. Kamiya, E. A. Davis, Jai Singh, "Percolative growth of photodarkening in amorphous As_2S_3 films," Physical Review B **62** (22), 14601–14604, 2000
[4]V. Lyubin, M. Klebanov, M. Mitkova, and T. Petkova, "Polarization-dependent, laser induced anisotroic photocystallization of some amorphous chalcogenide films," Appl. Phys. Lett. **71**, 2118(1997)
[5]Vladimir V. Poborchii, Alexander V. Kolobov, and Kazunobu Tanaka, "Photomelting of selenium at low temperature," Appl. Phys. Lett. **74**, 215 (1999)
[6]K. Takana, "Photoexpansion in As2Se3 glass," Phys. Rev. B **57**, 5163 (1998)
[7]P. Krecmer, A. M. Moulin, R. J. Stephenson, T. Rayment, M. E. Welland, S. R. Elliott, "Reversible nanocontraction and dilatation in a solid induced by polarized light," Science **277**, 1799 (1997).
[8]C. Y. Yang, M. A. Paesler and D. E. Sayers, "Measurement of local structural configurations associated with reversible photostructural changes in arsenic trisulfide films," Phys. Rev. B **36**, 9160 (1987).
[9]V. Kolobov, H. Oyanagi, Ke. Tanaka and K. Tanaka, "Structural study of amorphous selenium by in situ EXAFS: Observation of photoinduced bond alternation," Phys. Rev. B **55**, 726 (1997).

[10]M. Frumar, A. P. Firth and A. E. Owen, "Reversible photodarkening and structural changes in As_2S_3 thin films," Philos. Mag. B **50**, 463 (1984).

[11]L. F. Gladden, S. R. Elliott and G. N. Greaves, "Photostructural changes in bulk chalcogenide glasses: An EXAFS study," J. Non-Cryst. Solids **106**, 189 (1988).

[12]S. R. Elliott and A. V. Kolobov, "Photostructual changes in amorphous $As_{50}Se_{50}$ films: an EXAFS study," Philos. Mag. B **61**, 853 (1990).

[13]E. A. Stern, "Theory of the extened x-ray-absorption fine-structure," Phys. Rev. B **10**, 3027 (1974)

[14]J. Rehr, J., Albers, R.C., and Zabinsky, "High-order multiple-scattering calculations of x-ray-absorption fine structure," Phys. Rev. Lett. **69**, 3397 (1992)

[15]Renninger A. L. and Averbach B. L., "Cyrystalline structure of As_2Se_3 and As_4Se_4," Acta Cryst., Sec. B, **29**, 1583 (1973)

[16]Jun Li and D. A. Drabold, "First-principles molecular-dynamics study of glassy As_2Se_3," Phys. Rev. B **61**, 11998 (2000)

Electro-Optic and Ferroic Materials in Optoelectronic Applications

INVESTIGATIONS ON HIGH RESPONSE SPEED AND HIGH INDUCED STRAIN OF PHOTOSTRICTIVE DOPED PLZT CERAMICS

Patcharin Poosanaas-Burke
Manufacturing and Design Technology Center (MDTC)
The National Metal and Materials Technology Center (MTEC),
Bangkok 10400, Thailand

Bhanu Vetayanugul, Thanakorn Wasanapiarnpong, and Sutin Kuharuangrong
Department of Materials Science, Chulalongkorn University,
Bangkok 10330, Thailand

Kenji Uchino
International Center for Actuators and Transducers,
Materials Research Laboratory, PA 16801, USA.

ABSTRACT

A photostrictive actuator is a device in which light illumination induces motion. The phenomenon has greatest applications in the fields of micro-mechanics and optical communications. It is significant and promising for future micro-device technology. We have reported high photocurrent and high photovoltage observed in PLZT (lanthanum-modified lead zirconate titanate ceramics) 4/48/52 and PLZT 5/54/46, respectively. In this study the optimization of photostrictive PLZT ceramics through the 4/48/52 and 5/54/46 compositions with Nb_2O_5 and Gd_2O_3 dopants was studied in order to improve the response speed and induced strain.

INTRODUCTION

Photostriction is a phenomenon in which strain is induced in the sample by incident light. In principle, this effect arises from a superposition of the photovoltaic effect, i.e. generation of large voltage from the irradiation of light, and the converse-piezoelectric effect, i.e. expansion or contraction under the voltage applied. The figure of merit of photostriction may be expressed as the product of photovoltage, E_{ph}, and the piezoelectric constant,

d_{33}. Therefore, for application purposes, enhancement and/or optimization of photostrictive properties requires consideration of both terms in the figure of merit.

Because of relatively high piezoelectric coefficient, lanthanum-modified lead zirconate titanate (PLZT) ceramic is one of the most promising photostrictive materials for wireless photo-driven actuators application. A rigorous investigation on the composition dependence, especially near the morphotropic boundary (MPB), of photovoltaic properties in PLZT ceramics was carried out and the maximum photocurrent and photovoltage was reported for tetragonal phase 4/48/52 PLZT and for PLZT 5/54/46, respectively.[1] In addition the dopant ions, Nb_2O_5 and Gd_2O_3 (0.5 at% concentration), were found to enhance photovoltaic and photostrictive properties of PLZT ceramics.[2-4] However the optimization of photostrictive PLZT ceramics through the 4/48/52 and 5/54/46 compositions with Nb_2O_5 and Gd_2O_3 dopants has never been studied further. Since a material can be tailored through the composition and dopants depending on the requirements of a higher response speed, which is governed by photocurrent, or a larger magnitude of strain, which depends on photovoltage, the investigation of Nb_2O_5 and Gd_2O_3 dopants on PLZT (4/48/52) and (5/54/46) will be worthwhile for the fabrication of photostrictive devices with high response speed and high induced strain.

In this study, photostrictive PLZT 3/52/48, 4/48/52 and 5/54/46 ceramics with Nb_2O_5 and Gd_2O_3 dopants were fabricated by a conventional oxide mixing process in order to improve the response speed and induced strain.

EXPERIMENTAL PROCEDURE

PLZT (3/52/48) ceramics with 3 at% La and a Zr/Ti ratio of 52/48, PLZT (4/48/52) ceramics with 4 at% La and a Zr/Ti ratio of 48/52 and PLZT (5/54/46) ceramics with 5 at% La and a Zr/Ti ratio of 54/46 were chosen due to their high response speed and high photo-induced strain.[1] Nb_2O_5 and Gd_2O_3 were selected to be the dopants for PLZT ceramics. PLZT (3/52/48), (4/48/52) and (5/54/46) doped 0.5 at% Nb_2O_5 and Gd_2O_3 ceramics were synthesized by a conventional oxide mixing process. The calcining temperature was chosen at 950 °C for 10 hrs and sintered at 1250 °C for 2 hrs in a closed crucible. The sintered pellets were poled under 2 kV/mm at 100 °C for 30 min. Dielectric properties of PLZT samples were measured with an impedance analyzer (Hewlett Packard: 4192A LF Impedance Analyzer). Piezoelectric properties were measured using a Berlincourt Piezo d-meter Model CADT). The photovoltaic and photostrictive properties were done using the electrometer and displacement sensor under the irradiation of high-pressure mercury lamp (Ushio Optical Modulex H500) with intensity of 2.1 mW/cm^2. The details of the measurement setup are reported earlier.[5]

RESULTS AND DISCUSSION
Phase Analysis
The effect of dopant on the lattice parameters, tetragonality (c/a ratio), and unit cell volume of PLZT ceramics was observed and shown in Table I. It was found that Gd$_2$O$_3$ and Nb$_2$O$_5$ increased the unit cell volume as well as the lattice parameters of PLZT samples. The tetragonality showed the different between each composition however the effect of dopant on the tetragonality was unseen. PLZT 4/48/52 ceramics exhibited the highest tetragonality while the smallest values presented in PLZT 5/54/46 ceramics. In general the lattice parameters increased with increasing the Zr/Ti ratio while the c/a ratio decreased with increasing La content. PLZT 4/48/52 ceramic located in the tetragonal phase which is far from the morphotropic boundary (MPB), therefore it exhibited the highest c/a ratio, as compared to the other two compositions. PLZT 5/54/46 located closest to the MPB which confirmed by the lowest c/a ratio of PLZT in this composition.

Table I. Lattice parameters, tetragonality, and unit cell volume of undoped and doped PLZT ceramics calculated from the XRD patterns

Compositions	Dopants	c (Å)	a (Å)	c/a ratio	Unit cell volume (x10^{-23}cm^3)
3/52/48	Undoped	4.1303	4.0427	1.0217	6.7503
	Gd$_2$O$_3$	4.1592	4.0703	1.0218	6.8907
	Nb$_2$O$_5$	4.1592	4.0703	1.0218	6.8907
4/48/52	Undoped	4.1339	4.0359	1.0243	6.7335
	Gd$_2$O$_3$	4.1592	4.0634	1.0236	6.8673
	Nb$_2$O$_5$	4.1702	4.0738	1.0237	6.9208
5/54/46	Undoped	4.1592	4.0878	1.0175	6.9501
	Gd$_2$O$_3$	4.1739	4.1019	1.0176	7.0228
	Nb$_2$O$_5$	4.1592	4.0878	1.0175	6.9501

Microstructure Analysis

Micrographs of the thermally etched surfaces of PLZT ceramics were taken using Scanning Electron Microscopy. The average grain sizes calculated from the intercept method are plotted in Fig. 1.

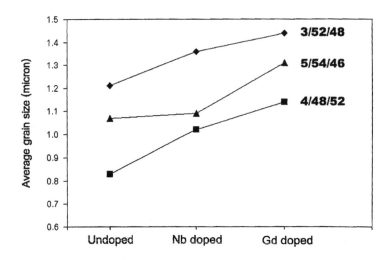

Fig.1. Average grain sizes of PLZT ceramics as functions of composition and dopant.

The average grain sizes were found to be different in each composition of PLZT ceramics. The smallest grain size was presented in PLZT 4/48/52 while the largest grain size was found in PLZT 3/52/48. In general, the average grain sizes of PLZT ceramics decreased with La addition and increased with Zr/Ti ratio. It was also reported that grain sizes increased as the composition of PLZT ceramics approaching the MPB region.[1,6,7] The results in this study showed the same tendency with the previous studies. PLZT 4/48/52 ceramics located in the tetragonal phase, far from the MPB region, thereby the smallest grain size was observed. PLZT 5/54/46 ceramics provided average grain sizes in between PLZT 3/52/48 and PLZT 4/48/52 ceramics. This is due to the higher percentage of La addition and the composition closest to the MPB region.

In addition, both Nb_2O_5 and Gd_2O_3 dopants increased grain sizes of PLZT ceramics in all the compositions. Gd_2O_3 was found to be more effective in increasing PLZT grain sizes compared to Nb_2O_5. The results were contradicted to the previous studies on the effect of dopants.[2,8] In general, donor dopants are associated with charged vacancies and impurity ions which

Optoelectronic Materials and Technology

resulted in decreasing the grain boundary mobility of PLZT ceramics. However it is possible that the impurity ions from this study didn't impede grain boundary mobility due to the lower amount of impurity ions or charged vacancies formed than expected

Dielectric and Piezoelectric Properties

Capacitance (C) and the dissipation factor of all the samples were measured by an Impedance Analyzer (Hewlett Packard 4192A LF) at 100 Hz. Room temperature dielectric constant of PLZT ceramics was calculated from capacitance values. Nb_2O_5 dopant was found to increase dielectric constant in all the compositions of PLZT ceramics. Among these three compositions, PLZT 5/54/46 showed the highest dielectric constant. The maximum dielectric constant was found in PLZT 5/54/46 doped with Nb_2O_5. The same tendency was also found in piezoelectric constant. High dielectric constant sample was found to have large piezoelectric constant. The maximum dielectric and piezoelectric constant were found at Nb_2O_5 doped PLZT 5/54/46 ceramic. The dielectric constant (K) and piezoelectric constant (d_{33}) are plotted as functions of composition and dopant in PLZT ceramics as shown in Fig. 2.

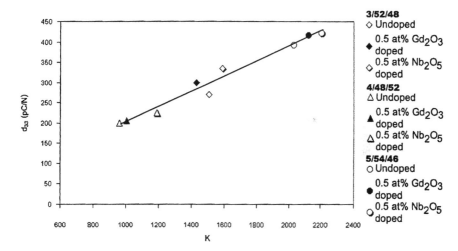

Fig. 2. The relation between dielectric constant (K) and piezoelectric constant (d_{33}) of poled PLZT ceramics.

Photovoltaic and Photostrictive Properties

A plot between the applied voltage and the measured current was used to calculate photocurrent and photovoltage. Photovoltage (E_{ph}) was determined from the intercept of the horizontal applied voltage axis while photocurrent (I_{ph}) was determined from the intercept of the vertical measured current axis. Photoconductance (G_{ph}) of the sample was calculated using the relation

$$E_{ph} = \frac{I_{ph}}{G_{ph}} \tag{1}$$

It was shown that both dopants decreased photocurrent in all compositions while they showed the significant effect in increasing photovoltage of PLZT ceramics. Composition was found to play an important role on photovoltage of PLZT ceramics. PLZT 5/54/46 provided the lowest photocurrent but the highest photovoltage. On the other hand, PLZT 4/48/52 provided the maximum photocurrent while the minimum was presented in PLZT 5/54/46. This finding agreed well with the previous study.[1] Nb_2O_5 was found to be the better dopant in enhancing photovoltage of PLZT ceramics compared to Gd_2O_3 dopant. PLZT 3/52/48, which was reported in many previous studies to have a good photostrictive property, showed the properties in between the other two compositions as shown in Fig.3.

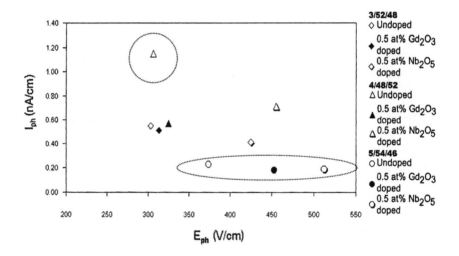

Fig. 3. Variation of photovoltaic properties with composition and dopant in PLZT ceramics (Light intensity 2.1 mW/cm^2).

As illustrated in Fig.3, it is clearly seen that the maximum photocurrent and photovoltage are obtained at different compositions of PLZT ceramics. In addition, Nb_2O_5 dopant only enhanced photovoltage in PLZT 5/54/46 while it decreased photocurrent in PLZT 4/48/52. This can be explained using the tetragonality of PLZT ceramics. Fig. 4 shows the relationship between photocurrent with the tetragonality of PLZT ceramics. It is clearly seen that the enhancement in photocurrent of PLZT ceramics was found in higher tetragonality samples while the stability in higher symmetry cubic phase (c/a raio =1) resulted in a decrease in photocurrent (Fig. 4).

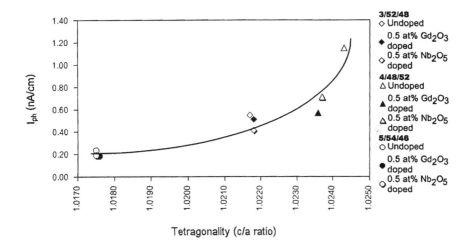

Fig. 4. Photocurrent measured under light intensity 2.1mW/cm^2 as a function of the tetragonality (c/a ratio) of PLZT ceramics.

In addition, photoconductance of PLZT ceramics was calculated and plotted as a function of photocurrent in Fig. 5. Higher photocurrent samples provided higher photoconductance. This is because photocurrent was strongly effected by material conductivity (on the other hand, resistivity). In general, higher conductivity materials will ease the flow of carrier or the generated photocurrent under illumination.

Photovoltage and photoconductance were also related. Photovoltage was inversely proportional to photoconductance as shown in Fig. 6 where photovoltage is plotted as a function of photoconductance.

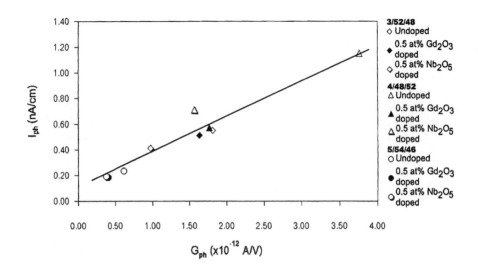

Fig. 5. Photocurrent (I_{ph}) as a function of photoconductivity (G_{ph}) in PLZT ceramics (Light intensity 2.1mW/cm^2).

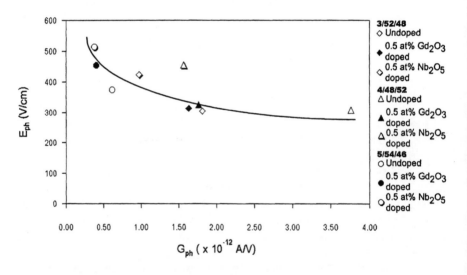

Fig. 6. Photovoltage (E_{ph}) as a function of photoconductance (G_{ph}) in PLZT ceramics (Light intensity 2.1mW/cm^2).

The figure of merit for response speed was reported in an earlier paper to be equal to "$d_{33}I_{ph}/C$" and the magnitude of strain is "d_{33} x E_{ph}".[1] Both of the merits were calculated. The merit of response speed was obtained at PLZT 4/48/52 ceramic due to its highest photocurrent, low capacitance and high c/a ratio. The merit of photo-induced strain was presented in PLZT 5/54/46 doped with Nb_2O_5, which displayed the highest values of dielectric constant, piezoelectric constant, and photovoltage. Due to the above results, PLZT 4/48/52 ceramic is suitable for high response speed applications, such as vibrators or photophone, while Nb_2O_5 doped PLZT 5/54/46 ceramics should be selected for high photo-induced strain applications.

CONCLUSIONS

Properties of photostrictive PLZT ceramics were investigated through the doping effect of Nb_2O_5 and Gd_2O_3 on PLZT 3/52/48, 4/48/52, and 5/54/46 compositions. It was found that 0.5 at% Nb_2O_5 doped PLZT 5/54/46 ceramic displayed the maximum dielectric constant, piezoelectric constant and photovoltage made it suitable for high displacement applications, such as photo-driven robot. On the other hand, undoped PLZT (4/48/52) ceramics exhibited the maximum photocurrent with high tetragonality and low dielectric constant made it suitable for high response speed applications, such as vibrators or photo-acoustic devices.

REFERENCES

[1]P. Poosanaas and K. Uchino, "Photostrictive effect in lanthanum-modified lead zirconate titanate ceramics near the morphotropic phase boundary," *Materials Chemistry and Physics,* **61**, 36-41 (1999).

[2]P. Poosanaas, K. Tonooka, I.R. Abothu, S. Komarneni, and K. Uchino, "Influence of Composition and Dopant on Photostriction in Lanthanum-Modified Lead Zirconate Titanate Ceramics," *J. Intelligent Material Systems and Structures,* **10**, 439-445 (1999).

[3]M. Taminura and K. Uchino, "Effect of impurity doping on photostriction in ferroelectric ceramics," *Sensors and Materials,* **1**, 47-56 (1988).

[4]S. Y. Chu and K. Uchino, "Effects of impurity doping on photostrictive effect in PLZT ceramics," *J. Advanced Performance Materials,* **1**, No. 2, 129-143 (1994).

[5]P. Poosanaas, A. Dogan, A. V. Prasadarao, S. Komarneni, and K. - Uchino, "Photostriction of sol-gel PLZT ceramics, " *J. Electroceramics,* **1**, No. 1, 105-111 (1997).

[6]Haertling, G., Proc. 9th IEEE Int'l Symposium on Applications of Ferroelectrics Transactions, 313 (1994).

[7]Akbas, M.A., Reaney, I.M., and Lee, W.E., *J. Mater. Res.,* **11,** 2293 (1996).

[8]Atkin, R.B., Holman, R.L., and Fulrath, R.M., *J. Amer. Ceram. Soc.,* **66,** 253 (1965).

SINGLE CRYSTAL ELECTRO-OPTIC FIBER IN OPTICAL WAVELENGTH SHIFT

Shilpi Bhargava and Ruyan Guo
Materials Research Institute and Department of Electrical Engineering
The Pennsylvania State University, University Park, PA 16802, USA

ABSTRACT

Optical wavelength shifting is observed in single crystal electrooptic fibers subjected to microwave field. LiNbO$_3$ and doped Sr$_{0.61}$Ba$_{0.39}$Nb$_2$O$_6$ single crystals grown by the laser heated pedestal growth technique were tested. This paper presents the preliminary results on the design and verification of the photon acceleration process to explore the material properties corresponding to desired applications.

INTRODUCTION

Optical modulation is becoming important in applications such as communications and image processing. The need for active devices at microwave wavelengths for radar and communication applications has led to the application of the electro-optic effect in crystals using modulation, switching, and frequency multiplexing.[1,2,3] The appreciable values and their electric field sensitive linear and non-linear dielectric properties in the visible spectrum are the main advantage of ferroelectrics materials. LiNbO$_3$ and LiTaO$_3$ have been used for phase shifting and amplitude modulation at millimeter wavelengths as have perovskites types such as BaTiO$_3$ and tungsten bronze materials.[4] A number of other devices proposed at millimeter wavelengths using the electro-optic or non-linear properties of ferroelectrics include tunable filters and electrically scanned antennas.

The induced phase shift by linear electrooptic effect for a transverse modulating field is given by

$$\Gamma = \frac{\pi L}{\lambda} n^3 r E \qquad (1)$$

where L is the crystal length, E is the applied field, n is the refractive index and r is the electro-optic coefficient.

Chirping or compression techniques refer to the method of generating short optical pulses, which cannot be obtained directly from the laser oscillator. The self-phase modulation by an optical fiber to produce a nearly linear frequency shift or chirp with subsequent compression techniques in a dispersive delay line is one example.[5] The technique is not applicable to low-power pulses since the self-phase modulation is a non-linear optical process that is intensity dependent. Various techniques have been described to show pulse compression techniques using electro-optic modulators to achieve frequency shift and chirp.[5]

An optical wave with Gaussian intensity traveling in an electro-optic medium has a traveling wave modulation given by [5]

$$E(x,t) = \frac{E_0}{2} \exp\left[-2\ln 2\left(\frac{t - x/v_g}{\tau_i}\right)\right] . \exp\{j[k_0 - \varpi_0 t + \Gamma \cos(k_m x - \varpi_0 t + \phi_0)]\} \quad (2)$$

where v_g is the optical velocity, k_0 and ω_0 are the free-space optical propagation constant and carrier frequency, respectively, k_m is the microwave propagation constant, and ϕ_0 is an adjustable phase shift. Γ, the modulation index or peak phase deviation, is dependent upon the electro-optic coefficient r_{ijk}, the interaction length L and electric field strength, as given in Eqn.(1).

If the input pulse width τ_i is small compared to the period, the phase modulation is approximately quadratic over the pulse envelope. The instantaneous frequency is then linearly chirped and pulse compression can be achieved by an appropriate dispersive delay line.

Earlier studies have been done with the longitudinal coupling of the microwave and optical signals as in [5] and [7]; where wavelength shift is achieved by modulation of the optical signal by microwave signal. The microwave signal is very slow compared to the optical signal. The essential principal applied is that the refractive index of a medium is a function of both space and time. The time varying effect of the electric field could give rise to a wavelength shift.[6] Previous work on this theory has been done with waveguides made of materials such as LiNbO$_3$ and Ti: LiNbO$_3$ phase modulator. [5,6] Both the studies have been performed with the optical beam and the microwave co-propagating in the same longitudinal direction.

In this work, we attempt to create a frequency shift with transverse microwave field application. This takes the principal of refractive index changing with time into consideration. The light pulse is incident along the length of the fiber. The pulse travels the length with a certain velocity as per the refractive index of the material. The microwave field is applied transverse to the crystal along the fiber. The electric field increases in intensity as the pulse travels and this gives rise to a change in refractive index with time. As long as the pulse remains within a slope of the electric field it experiences a linear increase in the refractive index. This will cause the pulse to shrink or expand depending on the slope of the electric field.

The concept is explained here with reference to Fig. 1. A slow moving electric field (microwave E-field) is coupled into an element, a single crystal fiber in this case, and there is a portion of the pulse waveform in the sample. A

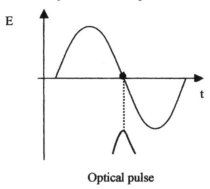

Fig.II.2. Pulse modulation/frequency chirp concept.

longitudinally coupled optical signal with a pulse width smaller than quarter of the E-field wavelength will experience a slope of refractive index changing by n ± Δn. One side, frontal or back, of the optical pulse experiences a larger n while the other a smaller one. This causes a chirp in the pulse and could give rise to pulse compression or a frequency shifted output.

One point to be considered here is the small diameter of the fiber across which the microwave signal travels and hence the extent of the microwave field distribution is considerably flat. This approach is conceptually simple but considerably practical in real device designs. The possibility of the experiment having a few drawbacks is not ruled out since a conventional waveguide type structure is not used. At this point, the attempt is made to verify the feasibility of such a principle and to explore materials-related issues for future device design.

EXPERIMENTAL PROCEDURE

The demonstration of microwave tuned optical wavelength shifting is carried out using a second harmonic Q-switched Nd:YAG laser as the light source. The microwave source is the HP4291A impedance analyzer, which has a power output of 1mW. The power is fed into a pair of coaxial cables, which feeds the signal into the HP4291A Test Head. A high temperature component fixture HP16194A, is the sample holder mounted on the test head. This sample holder is designed for surface mount devices, which is ideally suited for the test where the E-field is applied perpendicular to the optical wave propagation direction. For this purpose the samples have to be transparent in the visible region.

The laser, Quantronix 416, is a Q-switched pulsed high power green laser, which is used as the optical signal source. The power is attenuated to a few millwatts to avoid damage to the crystal. The optical pulses are observed in a high speed digital oscilloscope, capable of capturing upto 8GSa/s. The optical pulse is sensed by a photodiode, which feeds the signal to a preamplifier and then to the oscilloscope. The set-up is shown schematically in Fig. 2.

The oscilloscope used for capturing the pulses can capture up to 125 ps at a maximum frequency of 1.5 GHz. For our purposes this capture rate is sufficient since the pulses generated from the laser are typically 200-900ns. A repetition rate of about 3-6 kHz for the laser pulses was used.

At a higher pulse rate and smaller pulse widths we would require a spectrum

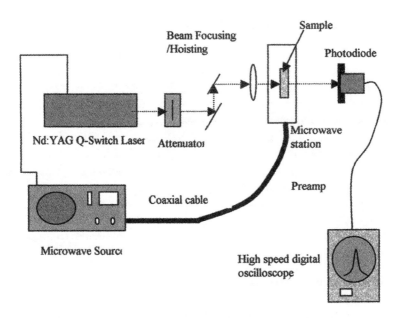

Fig.2. Schematic Set-up for the frequency modulation experiment.

analyzer or an autocorrelator to measure the pulse width and frequency. The frequency shift is measured by measuring the time between two pulses before and after application of the microwave energy. The characteristics of the laser are such that the pulse width increases with reducing the repetition rate and increasing power. The repetition rate is always a harmonic of the microwave frequency since the two signals have to be in phase.

The frequency shift determined by Riaziat et al [7] was given by the equation shown below subject to an electric field E generated by the microwave signal.

$$f = f_0 \cdot e^{\pi\left(d_{eff}/\lambda_m\right)n_0^2 \cdot r_c \cdot E} \tag{3}$$

where f_0 is the initial light frequency, f is the final light frequency, λ_m is the wavelength of the guided microwave signal, and d_{eff} is the effective interaction length.

Without measuring optical signal frequencies, according to Kohler [5], if the index of refraction has a traveling-wave modulation and the input pulse is small compared to the period, $T = 2\pi/\omega_m$ of the microwave field, then the cosine term (Equ. (2)), can be expanded and phase modulation can be derived. In the absence of velocity mismatch, the compressed or expanded pulse, τ_i can be written in the normalized form as a ratio

$$\tau_i/\tau_0 = \sqrt{1+\left[\left(A\pi^2/\ln 2\right)\left(\tau_i/T\right)^2\right]^2} \tag{4}$$

this equation was used in this research for analysis of obtained results.

In the frequency domain this result and derivation is equivalent of canceling the quadratic phase. The average power of the incident optical pulse train has to be held within several milliwatts because above a certain value, the waveguiding is impaired. The average power of the microwave is also kept in the level of the optical signal so that unwanted predominance of one signal does not impede coupling.

Cerium doped strontium barium niobate Ce:$Sr_{0.61}Ba_{0.39}Nb_2O_6$ (Ce:SBN) crystals polished optically were used as test samples. The samples were grown at Penn State using the LHPG [8] technique, along c-axis and a-axis directions. Their electrooptic properties were reported previously.[9,10]

RESULTS AND DISCUSSION

The optical frequency shift is demonstrated on chosen crystals. The Ce:SBN crystal, with microwave applied along c-axis and light propagating along a-axis, shows chirping in the pulse width. Though small at this point due to the short length (4 mm) of the fiber used, the frequency shift is evident as shown in Fig. 3.

The change in the pulse dimensions before and after application of the microwave field, is more than just intensity reductions due to the fiber presence. It can be seen in Fig. 3. that the shift is about ~100 ns for a repetition rate of 10 kHz. The chirp in the pulse is due to the refractive index gradient established in the fiber because of the microwave signal. A repetition rate of 10 kHz is a 1000th harmonic of the microwave signal.

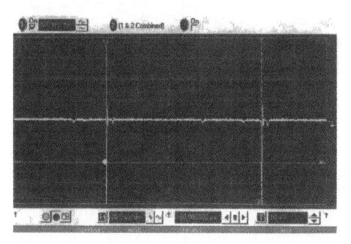

Fig.3. (a) Before applying electric field. The pulse on an expanded scale gives a width of approx. 500 ns. Sampling rate is 100MSa/s.

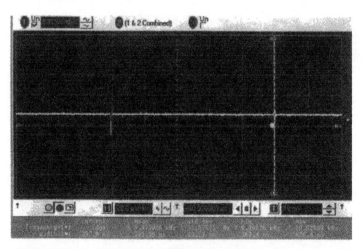

Fig. 3 (b) On applying electric field. The pulse on an expanded scale gives a width of approx. 400 ns. Sampling rate is 100MSa/s.

Optoelectronic Materials and Technology

An estimation of the frequency shifts possible with this type of configuration based on Eqn. (2) is given in Table I. This tabulation shows the dependence of the material on its electro-optic properties (r_c), dielectric constant (ε_r), the wavelength of the optical signal, microwave power and the sample size. It is practical to have a small sample size such that the device would be small too. Therefore, a high dielectric constant material with appreciable values of electro-optic coefficient would be suitable for achieving maximum frequency shift.

Table I. Estimated frequency shifts with laser power of ~20mW, sample cross-section of about 1mm², and a reduced pulse width of ~100ns.

S. No.	ε_r	r_c ($*10^{-12}$ m/V)	d_{eff}/λ_m	Δf (kHz)
1.	100	50	0.25	~0.2
2.	200	75	0.10	~1.81
3.	500	150	0.05	~3.2
4.	1000	200	0.25	~17.5
5.	2000	250	0.10	~10.48
6.	5000	400	0.05	~10.54

CONCLUSION

Wavelength shift or frequency chirp has applications in communications and scientific applications such as spectroscopy, wavelength multiplexing, pulsed transmission systems and optical systems. Utilizing the electro-optic properties of ferroelectric tungsten bronze SBN single crystals grown by a laser heated pedestal growth (LHPG) technique, a wavelength shifter with crystalline electro-optical material acting as a wave-guide that couples both the optical and the microwave power into the device is demonstrated. Interactions among material parameters and the frequency tuning are analyzed in the preliminary work. Single crystal fibers grown by the LHPG technique have shown potential in such transverse microwave tuned optical applications such as converting a fixed frequency pulsed laser into a widely tunable one.

There are still many issues left for further investigation requiring a more detailed analysis of the transverse coupling of the microwave and optical signal. It can be concluded however that the transverse microwave field configuration is easier to be adapted in communication applications. By extending the same principles, this arrangement may be applied to the commercial optical wavelength 1.55 μm.

ACKNOWLEDGEMENT

The authors thank Prof. Amar Bhalla, Penn State University, for stimulating discussions and interest on the subject. The financial support of US National Science Foundation under grant #0075917 is gratefully acknowledged.

REFERENCES

[1] Alan Eli Willner, "Mining the optical bandwidth for a terabit per second", IEE Spectrum, p. 32 (April 1997).

[2] M. S. Borella, J. P. Jue, D. Banerjee, B. Ramanmurthy and B. Mukherjee, "Optica components for WDM lightwave networks," Proceedings of the IEEE **85**, 127 (1997)

[3] J. N. Eckstein, A. I. Ferguson, and T.W. Hansch, "High resolution two photo spectroscopy with picosecond light pulses," Phys. Rev. Lett. **40**, 847 (1978).

[4] Klein M.B., "Ferroelectric materials for Electro-optic Devices at Millimeto Wavelengths", *Ferroelectrics*, Vol. 50, pp. 301-306 (1983).

[5] Kohner B,H., "Active pulse compression using an integrated electro-optic phas modulator", *Appl. Phys. Lett.*, Vol. **52**, No.14, pp. 1122-1124 (1988).

[6] Ho W.W., Hall W.F., Neurgaonkar R.R., "Dielectric Properties of Ferroelectr Tungsten Bronze $Ba_{2-x}Sr_xK_{1-y}Na_yNb_5O_{15}$ crystals at RF and millimeter wav frequencies", *Ferroelectrics*, Vol. **50**, pp. 325-330 (1983).

[7] Riaziat M.L., Virshup G.F., Eckstein J.N., "Optical Wavelength Shifting b Traveling-Wave Electro-optic Modulation", *IEEE Photonics Technology Letter* Vol. **5**, No. 9 (1993).

[8] J. Yamamoto and A.S. Bhalla, "The Growth and Characterization of Tungste Bronze $Sr_xBa_{1-x}Nb_2O_6$ Single Crystal Fibers", *Materials Research Bull.* **24**, 76 (1989)

[9] J. K. Yamamoto and A.S. Bhalla, "Strntium barium niobate single crystal fiber optical and electrooptic properties," *J. Appl. Phys.* **70**, 3215 (1991).

[10] Guo R., Wang J.F., Povoa J.M., Bhalla A.S., "Electrooptic properties and the temperature dependence in single crystals of Lead Barium Niobate and Strontiu Barium Niobate", *Materials Letters*, **42**, pp.130-135 (2000).

FABRICATION OF PHOTONIC BANDGAP STRUCTURES BY FUSED DEPOSITION OF MULTIMATERIALS

M.E. Pilleux
Universidad de Chile - IDIEM
Plaza Ercilla 883
Santiago, Chile

Y. Chen and Y. Lu
Rutgers University
Electrical & Computer Eng. Dept.
Piscataway, NJ 08854

E. Niver
New Jersey Institute of Technology
Electrical & Computer Eng. Dept.
Newark, NJ 07102-1982

M. Allahverdi, E.K. Akdogan, and
A. Safari
Rutgers University
Ceramic & Materials Eng. Dept.
Piscataway, NJ 08854-8085

ABSTRACT

The fabrication of 1-D and 3-D photonic bandgap (PBG) structures was achieved using the Fused Deposition of Multimaterials (FDMM) technique with Transtech D-50 and alumina as high permittivity materials. After the structures were fabricated using the FDMM technique, they were subjected to a binder burnout cycle to remove the polymer matrix and then to a presintering and sintering stage for densification. Complex parts were made. The electromagnetic measurements made on the 3D alumina log-pile type structures indicated a precise match with the bandgap values predicted using two modeling techniques (16-24 GHz). A similar procedure was followed with the 1-D PBG structures. FDMM has proven to be an efficient and flexible tool for rapid prototyping of PBG structures.

INTRODUCTION

Extensive theoretical and experimental work has been devoted to photonic bandgap (PBG) crystals in order to understand and exploit the properties of these structures in various dimensions and many materials [1-3]. The main feature of these crystals is their capability to affect the radiative dynamics within the structure so no electromagnetic modes are available within the dielectric. A variety of applications are possible, such as thresholdless lasers, high quality-single mode LEDs, microwave antennas, light diodes, an all kinds of optical circuits have been suggested, and some have already been demonstrated [4]. For the microwave-millimeter wave region, in which our interest is focused, applications involve control of signal propagation, quiet oscillators, frequency selective surfaces, narrow band filters, and antenna substrates. There have been many reports for the application of PBG structures as antenna substrates [5,6].

Research in photonic structures using alumina as the high permittivity dielectric material has been performed by many researchers [7-10]. The use of anodic porous alumina formed by anodization of aluminum in an appropriate acid solution has attracted attention as a starting material for 2-D PBG structures with typical dimensions in the nano- or micrometers [7-9]. Also, Feiertag et al. developed a microfabrication technique for building 3-D PBG structures using x-ray lithography with bandgaps in the infrared region [10]. Jin et al. made microwave measurements on a 2-D octagonal quasiperiodic photonic crystal made of an array of 23 x 23 rows of alumina cylinders [11]. The authors measured a bandgap between 8.9 and 10.5 GHz, they saw that the position and width of the bandgap did not depend on the incidence direction, and that it can appear even if the array's dimensions are lowered to 11 rows of cylinders. The authors also fabricated waveguides with this array, demonstrating the efficiency of straight and bending guides.

Our research has focused on the design of PBG structures using alumina (3-D) and Transtech D-50 (1-D) as a low-loss material with a higher permittivity material than alumina. To our knowledge, there is no report of the use of Transtech D-50 for this application. Fused Deposition of Ceramics (FDC), or its variation for multimaterials (FDMM), presents several advantages for the fabrication of the PBG structures for use in the microwave region, since the minimum part shapes are in the order of 0.5 mm. The main advantage of this technique is the rapid prototyping of the complex design. While other fabrication processes make bulk pieces of the dielectric material and then cutting, drilling and/or etching is required to remove the bulk materials in order to fabricate the alternating high and low dielectric constant materials, FDMM only deposits the desired material in the fabrication procedure of the structure and is also able to fabricate PBG devices that cannot be made otherwise in a single fabrication process.

EXPERIMENTAL PROCEDURE
Modeling of Alumina Structures

The modeling of the alumina structures was based on a finite element shareware Fortran program written by the Photonics Research Team of Imperial College (London, UK), which was adapted to this research in order to incorporate the specific structures that were modeled [12,13]. Frequency-domain modeling was also performed using the High Frequency Structure Simulator (HFSS, Ansoft Corporation, Pittsburgh, PA).

If it is assumed that the solution for the magnetic field in Maxwell's equations is a superposition of plane waves, these equations can be reduced to a system of finite difference equations, which can be solved using standard numerical methods. As previously indicated, our simulation used two different computational methods to investigate the PBG structure. The first one, the Fortran-code software, was used to reformulate Maxwell's equations on a lattice by dividing the space into a set of small cells with a coupling between neighboring ones [12]. Then, it calculated the propagation of the EM fields through a periodic dielectric structure in a layer-by-layer manner by means of a transfer matrix [13]. The transfer matrix discretices Maxwell's equations in a simple cubic lattice.

Our second approach was to solve Maxwell's equations in the frequency-domain using the High Frequency Structure Simulator (HFSS) software. The wave-guide simulator method was used to calculate the EM wave distribution in the propagation direction (z-direction). The unit-cell concept is also applied in the simulation so that the structure is assumed to repeat in the x-y plane. The inputs for the program were the geometry of the structure, which is defined in terms of its unit cell, the monochromatic source and the boundary conditions at the surface edges.

The structure geometries that we have modeled with these approaches are shown in Table I. The size of the unit cell and, in consequence, the spacing between bars and the geometry of the cross-section of each bar of the structure, were chosen so that the bandgap would lie in the 15-95 GHz frequency range. Alumina was used as the high permittivity dielectric material (relative permittivity, $\varepsilon_r=9.6$) and air as the low permittivity one.

FDMM Fabrication

The fabrication of the PBG structures was performed using the multimaterial deposition equipment designed and fabricated at Rutgers University, which is described elsewhere [14]. The feedstock materials for the FDMM process were ceramic-loaded polymer composite filaments made of ICW-06 wax (Stratasys, Inc., Eden Prairie, MN) and A152-SG alumina

(Alcoa, New Milford, CT) or Transtech D-50 (Trans-Tech, Adamstown, MD) which is a powder of barium-titanium oxides. The filaments were fabricated by coating the powders with a surfactant by mixing 150 g of the ceramic in a 30 g/L solution of stearic acid in toluene and then mixing the slurry for 4 h. The mixture was then filtered in order to remove the solvent.. Once the coated powder was dried, it was mixed with ECG-9 composition thermoplastic binder (developed at Rutgers University [15]) in a Haake System 9000 high-shear mixer (Haake-Fisons, Paramus, NJ) with a twin-roller blade mixing bowl operating at 100 rpm. The powder volume fractions used for alumina were 60 and 62 vol.% solids loading, while for Transtech D-50 it was 56 vol.% solids. The compounded ceramic-binder system was then extruded at 90 °C into continuous filaments several meters long through a 1.78-mm diameter nozzle using the same system but with a single screw extruding attachment.

Table I. Simulation prediction of PBG structures.

PBG Structure	Diameter or side dimension (mm)	Pitch (mm)	Filling ratio	Bandgap (GHz)	$\Delta\omega_{gap}/\omega_{midgap}$
Cylindrical alumina rods in air	0.625	1.25	0.39	55-75	30%
Cylindrical air rods in alumina	0.625	1.25	0.39	48-57	17%
Cylindrical alumina rods in air	0.521	2.08	0.19	69-95	32%
Square alumina rods in air	2	8	0.25	16-23	36%
Square alumina rods in air	1	4	0.25	32-46	36%
Square alumina rods in air	3	8	0.375	14-19	30%
Square air rods in alumina rods	3	4	0.75	32-46	36%

In the case of alumina, the input from the modeled PBG structures was used as input for making a CAD file in order to fabricate one of the structures modeled with the FDMM equipment (see Figure 1). The structure was constituted of bars stacked in an alternating manner, parallel to each other in each layer, and perpendicular to the direction of the immediate neighboring layers. For every second layer, there is a shift in the position of the rods by half lattice constant . The geometry of this PBG structure required the use of a supporting material below the alumina rods so that the thermal treatments would not deform them, so ICW-06 wax was chosen for this purpose. The part was fabricated by the successive deposition of the wax and the alumina-loaded filament in a layer-by-layer manner. The alumina filament was extruded through a 500 μm diameter nozzle with a liquefier heated to 130 °C. The wax filament was extruded through a similar liquefier heated to 72 °C and with a similar diameter nozzle. The CAD/CAM system instructed the FDMM equipment to move the liquefiers in predefined tool paths.

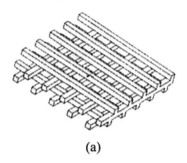

(a)

(b)

Figure 1. (a) CAD drawing of the unit cell of the PBG structure fabricated by FDMM. Each alumina bar is 28 mm long, with a 2 x 2 mm^2 square cross-section, and a pitch separation of 8 mm between centers. (b) Structure fabricated by FDMM after sintering at 1600 °C for 1 h..

To avoid the bending of the alumina bars while performing the binder-burn-out (BBO) process, the prior removal of the wax from the fabricated structure was necessary so it could be replaced by a temperature resistant support material for the BBO process. The wax removal was carried out in a furnace by placing the as-fabricated part for 10 min at a temperature of 110 °C. This time-temperature combination allowed the removal of the wax without deforming the alumina-polymer structure. The dewaxed structure was subsequently filled with zirconia powder in order to provide support for the overhanging alumina bars. This procedure proved effective for supporting the bars while not reacting with the structure. The removal of the zirconia was simple and was done using a flow of pressurized air. The BBO cycle procedure used was the same one as for the fabrication of other electr/oceramic materials made with this method [16]. The BBO was carried out by heating the structure to 550 °C for 1 h, immediately followed by a partial sintering at 1050 °C for 1 h. The heating from 100 to 550 °C was carried out at 10 °C/h in order to avoid any excessive degassing that might structurally affect the structure of the sample due to the calcination of the organic components. Finally, sintering was carried out at 1600 °C for 1 h in a different furnace. The sintering cycle densified the structure, leaving the finished structure shown in Figure 1(b).

The Transtech D-50 structures were designed as a 1D PGB with equally spaced cylindrical holes, on a slab of the material (see Figure 2), which will later be filled with barium titanate, electroded, and wired for the application. The BBO and sintering of these structures was much simpler than that of the alumina, since, as can be seen in Figure 2, the simple geometry allows it. However, the treatment times and temperatures were the same, except for the sintering temperature which was 1350 °C for Transtech D-50.

The electromagnetic measurements of the alumina structures were performed in a Network Analyzer (Hewlett Packard, model HP8510C) with open ended waveguides.

RESULTS AND DISCUSSION
Simulation of Alumina Structures
    The Transfer Matrix method was applied to calculate the energy bandgap of the alumina structure with rectangular and cylindrical rods as shown in Table I and Figure 3. We demonstrated that the filling ratio (i.e., the ratio between the volume of material in the unit cell and the total volume of the unit cell) and the dielectric constant ratios are the major factors affecting the bandgap existence and the width of the bandgap frequency. The lattice constants of the structure determine the starting frequency and the width of the bandgap. The shape of the dielectric rod is not important,

and the rod can be either of a high permittivity material surrounded by air or it can be made of air rods embedded in a dielectric material. The frequency of the bandgap scales linearly with the unit cell length, which is defined by the size and the space between the rods. This is due to the linearity of Maxwell's equations.

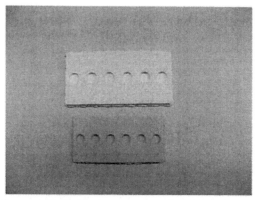

Figure 2. Transtech D-50 structures as-built by the FDMM equipment (top) and after the complete firing cycle (bottom). A shrinkage of 14.8% was observed.

With our second approach, using the High Frequency Structure Simulator (HFSS) software, the structure exhibited a bandgap starting around 14.7 GHz with a bandgap width of 8 GHz. Figure 4(a) shows the wave distribution at a frequency below the bandgap (12 GHz) and that the structure behaves as a homogeneous dielectric material with an effective permittivity between that of air and alumina. Figure 4(b) shows that, at frequencies inside the bandgap (16 GHz), the material behaves like a Bragg reflector. The transmission coefficient calculated by the HFSS program has the almost the same bandgap range as that shown in Figure 3 using the T-matrix approach.

Electromagnetic Measurements

The electromagnetic measurements of the alumina PBG was carried out on a stack of 4 "unit cells" of the sintered PBG structures. Each unit cell has 4 stacks of bars, so the measurement involved 16 layers of bars with the incident radiation perpendicular to the top side of the unit cell shown in Figure 1. A stop band was detected between 17.1 and 23.3 GHz with a maximum loss of 20 dB, as shown in Figure 5. These results are in excellent agreement with the simulation results for this structure shown in

Table I. However, since the measurements carried along the z-axis of the structure, and the x-y plane consisted of 6 layers of bars, the transmission loss is not as large as expected from the simulation (~40 dB).

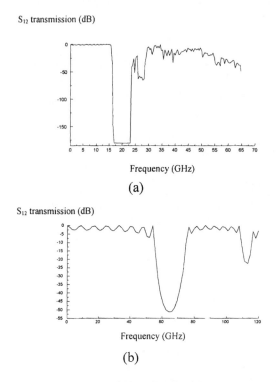

Figure 3. Transmission loss of the alumina structures (a) square cross-section rods (2 x 2 mm^2) with half-period shift (the transmission loss is calculated from the reflection coefficients, which approach to zero inside the bandgap) and (b) cylindrical rods (0.625 mm diameter).

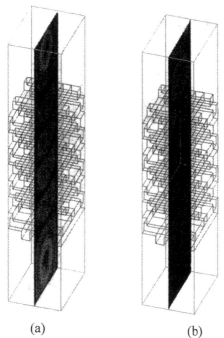

(a)   (b)

Figure 4. Electric field distribution in the propagation direction of the PBG structure made of rectangular alumina rods of 2 x 2 mm^2 cross section, with a pitch separation of 8 mm between rods, at (a) 12 GHz (below the bandgap) and (b) 16 GHz (inside the bandgap).

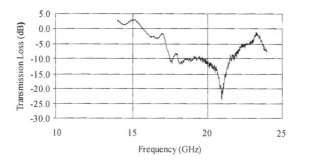

Figure 5. Electromagnetic characteristic of a PBG structure composed of 4 "unit cells" (geometry is shown in Figure 1). In the transmission versus frequency graph, a stop band is found between 17.1 and 23.3 GHz with a maximum loss of 20 dB.

CONCLUSIONS

Photonic bandgap (PBG) structures were designed and modeled in order to have a bandgap in the microwave frequency region. Successful fabrication of structures of alumina and Transtech D-50 was achieved. Computer simulation was performed using the T-Matrix and time-domain approaches on the alumina structures, resulting in the appearance of a bandgap in the expected microwave frequency region. Electromagnetic measurements in these structures confirmed the existence of the bandgap in the predicted frequency region. The modeling demonstrated that the photonic bandgap can be predicted in a structure of a given geometry and material, thus allowing the engineering of PBG structures for specific applications.

ACKNOWLEDGEMENTS

The New Jersey Commission of Science and Technology has sponsored this project under the Research Excellence Program.

REFERENCES

[1] J. P. Dowling, H. Everitt, and E. Yablonovitch, "Photonic & Sonic Band-Gap Bibliography", *http://home.earthlink.net/~jpdowling/ pbgbib.html* (2000). Continuously updated web page with comprehensive list of references on the subject.

[2] E. Yablonovitch, "Inhibited spontaneous emission in solid state physics and electronics," *Physical Review Letters*, **58**, 2059-2062 (1987).

[3] E. Yablonovitch, "Photonic band-gap structures," *Journal of the Optical Society of America*, **10** [2] 283-295 (1993).

[4] J. Joannopoulos, R. D. Meade, and J.N. Winn, "Photonic Crystals". Princeton University Press, Princeton, New Jersey, 1995.

[5] E. R. Brown and O. B. McMahon, "High zenithal directivity from a dipole antenna on a photonic crystal," *Applied Physics Letters*, **68** [9] 1300-1302 (1996).

[6] D. Sievenpiper, L. Zhang, R.F. J. Broas, N.G. Alexopolous, and E. Yablonovitch, "High-Impedance Electromagnetic Surface with a Forbidden Band," *IEEE Transactions on Microwave Theory and Technology*, **47** [11] 2059-2074 (1999).

[7] H. Masuda, M. Ohya. H. Asoh, M. Nakao, M. Nohtomi, and T. Tamamura, "Photonic crystal using anodic porous alumina," *Japanese Journal of Applied Physics, Part 2*, **38** [12A] L1403-L1405 (1999).

[8]O. Jessensky, F. Muller, and U. Gosele, "Self-organized formation of hexagonal pore arrays in anodic alumina," *Applied Physics Letters*, **72** [10] 1173-1175 (1998).

[9]S. Shingubara, O. Okino, Y. Sayama, H. Sakaue, and T. Takahagi, "Ordered two-dimensional nanowire array formation using self-organized nanoholes of anodically oxidized aluminum," *Japanese Journal of Applied Physics, Part 1*, **36** [12B] 7791-7795 (1997).

[10]G. Feiertag, W. Ehrfeld, H. Freimuth, H. Kolle, H. Lehr, M. Schmidt, M.M. Sigalas, C.M. Soukoulis, G. Kiriakidis, T. Pedersen, J. Kuhl, and W. Koenig, "Fabrication of photonic crystals by deep x-ray lithography," *Applied Physics Letters*, **71** [11] 1441-1443 (1997).

[11]C. Jin, B. Cheng, B. man, Z. Li, D. Zhang, S. Ban, and B. Sun, "Band gap and wave guiding effect in a quasiperiodic photonic crystal," *Applied Physics Letters*, **75** [13] 1848-1850 (1999).

[12]J. B. Pendry and A. MacKinnon, "Calculation of photon dispersion relations" *Physical Review Letters*, **69** [19] 2772-2775 (1992).

[13]J.B. Pendry, "Calculating photonic bandgap structure," *Journal of Physics: Condensed Matter*, **8**, 1085-1108 (1996).

[14]M.A. Jafari, W. Han, F. Mohammadi, A. Safari, S.C. Danforth, and N. Langrana, "A novel system for fused deposition of advanced multiple ceramics," *Rapid Prototyping Journal*, **6** [3] 161-174 (2000).

[15]T.F. McNulty, F. Mohammadi, A. Bandyopadghyay, D.J. Shanefield, S.C. Danforth, and A. Safari, "Development of a binder formulation for fused deposition of ceramics," *Rapid Prototyping Journal*, **4** [4] 144-150 (1998).

[16]T.F. McNulty, D.J. Shanefield, S.C. Danforth, and A. Safari, "Dispersion of lead zirconate titanate for fused deposition of ceramics," *Journal of the American Ceramic Society*, **82** [7] 1757-1760 (1998).

TWO-DIMENSIONAL MODELING OF GAUSSIAN BEAM PROPAGATION THROUGH AN ANISOTROPIC MEDIUM

Wook Lee and Ruyan Guo
Pennsylvania State University
University Park, PA 16802

ABSTRACT

Lasers having Gaussian profiles are commonly used as input light source in real applications that has properties inevitably different from the plane-wave solutions. Plane-wave approximation works well for cases where light source is distant from specimen; however, becomes deteriorating when the testing system is more compact. In this report, electromagnetic propagation of a Gaussian beam through an anisotropic medium is studied. The Gaussian beam is modeled as a sum of plane waves and an extended Jones matrix method is employed. Numerical Results are also presented for illustration.

INTRODUCTION

Jones matrix method [1] has been widely used to analyze the transmission performance of a birefringent network system as well as an individual optical element such as polarizers, retardation plates, liquid crystals, and so on. Although this 2 × 2 matrix representation is powerful and easily applicable, it is valid for the normal incidence case and cannot explain reflection losses. Many attempts to generalize the conventional Jones matrix formulation have been tried; among which a 4 × 4 matrix method [2,3] was applied to obtain an exact solution for the problem of off-axis light incidence to arbitrary birefringent media at the cost of quite a complicated calculation. If multiple reflections between the interfaces of the anisotropic medium can be neglected, a new 2 × 2 matrix method known as the extended Jones matrix method [4,5] can be applied, which is much easier to calculate, and also takes into account the effects of Fresnel refraction and reflection at the interfaces.

However, all these approaches are based on the plane-wave incidence and cannot be used for the practical applications where a laser is utilized as an input light source without extensive collimation. Therefore, a new treatment is desirable which considers the effects of the beam profile of the incident light. An angular spectrum of plane-wave technique [6] is employed for this purpose. A laser is modeled as a linearly polarized Gaussian beam and is expanded as a sum of plane waves having specific amplitudes and polarization states. Combined with the extended Jones matrix method, the transmission characteristics of a laser light through uniaxially anisotropic media having arbitrary crystal orientations are analyzed.

THEORY

Suppose that a Gaussian Beam is incident on a uniaxially anisotropic medium as shown in Fig. 1. The plate has a thickness of d and its input surface is located at $z = 0$. The incident beam propagates to the $+z$ direction and is polarized in the xz plane, i.e., $\mathbf{E}^{inc}(x,y,z) = E_x^{inc}(x,y,z)\hat{\mathbf{x}} + E_z^{inc}(x,y,z)\hat{\mathbf{z}}$. The time-dependent term $\exp(i\omega t)$ will be left out hereinafter. The focal point of the Gaussian beam can be arbitrarily chosen to be placed at $z = z_o$ without losing generality since the cross-sectional area of the medium is assumed infinitely large compared with that of the beam. The incident beam is then modeled such that the amplitude of the transverse electric field $E_x^{inc}(x,y,z)$ has a lowest-mode Gaussian distribution, and at the beam waist ($z = z_o$) it is written as

$$E_x^{inc}(x,y,z_o) = E_o \exp\left[\frac{-(x^2 + y^2)}{W_o^2}\right], \tag{1}$$

where E_o and W_o are the peak amplitude and minimum spot size, respectively. Following the plane-wave spectrum method introduced in Ref. 6, the total electric field of the incident Gaussian beam can be expressed in the discrete form as

$$\mathbf{E}^{inc}(x,y,z) \approx \sum_i \sum_j \mathbf{E}_{ij}^{inc}(x,y,z) = \sum_i \sum_j Amp(\xi_i, \zeta_j) \mathbf{Pol}(\xi_i, \zeta_j) \exp(-i\mathbf{k}_{ij} \cdot \mathbf{r}). \tag{2}$$

It can be seen from Eq. (2) that the incident Gaussian beam is represented as a sum of plane waves and each ijth plane wave $\mathbf{E}_{ij}^{inc}(x,y,z)$ is characterized with a specific amplitude $Amp(\xi_i, \zeta_j)$, polarization vector $\mathbf{Pol}(\xi_i, \zeta_j)$, and wave vector $\mathbf{k}_{ij}$, which are given by

$$Amp(\xi_i, \zeta_j) = p^2\left(\frac{E_o k^2 W_o^2}{4\pi}\right)\sin\xi_i \sin\zeta_j \exp\left(ik\sqrt{1 - u_{ij}^2}\, z_o\right)\exp\left(-\frac{W_o^2 k^2 u_{ij}^2}{4}\right), \tag{3}$$

$$\mathbf{Pol}(\xi_i, \zeta_j) = \hat{\mathbf{x}} - \frac{\cos\xi_i}{\sqrt{1 - u_{ij}^2}}\hat{\mathbf{z}}, \tag{4}$$

$$\mathbf{k}_{ij} = k\hat{\mathbf{u}}_{ij} = k\left(\cos\xi_i\hat{\mathbf{x}} + \cos\zeta_j\hat{\mathbf{y}} + \sqrt{1 - u_{ij}^2}\,\hat{\mathbf{z}}\right), \tag{5}$$

with

$$\mathbf{r} = x\hat{\mathbf{x}} + y\hat{\mathbf{y}} + z\hat{\mathbf{z}}, \tag{6}$$

$$u_{ij} = \sqrt{\cos^2\xi_i + \cos^2\zeta_j}. \tag{7}$$

Here $k = |\mathbf{k}_{ij}| = 2\pi/\lambda$ and λ is the wavelength outside of the uniaxial medium in free space. $\mathbf{r}$ is the position vector and $\hat{\mathbf{u}}$ is the unit wave vector of the ijth plane wave which is defined in terms of the angles of direction cosines, ξ_i and ζ_j as shown in Fig. 2. When $u_{ij}^2 = \cos^2\xi_i + \cos^2\zeta_j > 1$, a part of plane waves decay exponentially with z, and their contributions to the total electric field is negligible for $W > \lambda$. Therefore, it is practically legitimate that nonevanescent plane waves that satisfy $u_{ij}^2 > 1$ are only included in Eq. (2), and the summation is taken with a uniform interval width $p = \xi_{i+1} - \xi_i = \zeta_{j+1} - \zeta_j$ over the region where $0 < \xi, \zeta < \pi$.

The ijth plane wave $\mathbf{E}_{ij}^{inc}(x, y, z)$ in Eq. (2) can be decomposed into s (TE) and p (TM) waves according to the polarization state. Referring to Fig. 3, the s-polarized wave is defined such that the electric field of the plane wave is normal to the plane of incidence that is formed by two orthogonal unit vectors, $\hat{\mathbf{g}}_{ij} = (\cos\xi_i\hat{\mathbf{x}} + \cos\zeta_j\hat{\mathbf{y}})/u_{ij}$ and $\hat{\mathbf{z}}$. In the p-polarized wave, the electric field is parallel to the plane of incidence. Then, Eq. (4) can be rewritten as

$$\mathbf{Pol}(\xi_i, \zeta_j) = \frac{\cos\zeta_j}{u_{ij}}\hat{\mathbf{s}}_{ij} + \frac{\cos\xi_i}{u_{ij}\sqrt{1 - u_{ij}^2}}\hat{\mathbf{p}}_{ij}, \tag{8}$$

where $\hat{\mathbf{s}}_{ij}$ and $\hat{\mathbf{p}}_{ij}$ are, respectively, the normalized s and p polarization vectors of the ijth plane wave in which $\hat{\mathbf{s}}_{ij} \cdot \hat{\mathbf{s}}_{ij} = \hat{\mathbf{p}}_{ij} \cdot \hat{\mathbf{p}}_{ij} = 1$, and are given by

$$\hat{\mathbf{s}}_{ij} = \hat{\mathbf{g}}_{ij} \times \hat{\mathbf{z}} = \frac{1}{u_{ij}}(\cos\zeta_j\hat{\mathbf{x}} - \cos\xi_i\hat{\mathbf{y}}), \tag{9}$$

$$\hat{\mathbf{p}}_{ij} = \hat{\mathbf{u}}_{ij} \times \hat{\mathbf{s}}_{ij} = \frac{1}{u_{ij}}\left(\sqrt{1-u_{ij}^2}\cos\xi_i\hat{\mathbf{x}} + \sqrt{1-u_{ij}^2}\cos\zeta_j\hat{\mathbf{y}} - u_{ij}^2\hat{\mathbf{z}}\right). \qquad (10)$$

Note that in the case of normal incidence (i.e., $u_{ij} = 0$), the plane wave is linearly polarized along the x direction and can be arbitrarily defined as an s or p wave because the characteristics of both waves become identical. Using Eq. (8), Eq. (2) can be expressed as

$$\mathbf{E}^{inc}(x,y,z) = \sum_i\sum_j \mathbf{E}_{ij}^{inc}(x,y,z) = \sum_i\sum_j \left(A_{s,ij}\hat{\mathbf{s}}_{ij} + A_{p,ij}\hat{\mathbf{p}}_{ij}\right)\exp\left(-i\mathbf{k}_{ij}\cdot\mathbf{r}\right), \quad (11)$$

where $A_{s,ij}$ and $A_{p,ij}$ are amplitudes of the incident s and p waves for the ijth plane wave, which are given in the form of a column vector by

$$\begin{bmatrix} A_{s,ij} \\ A_{p,ij} \end{bmatrix} = p^2\frac{E_o k^2 W^2 \sin\xi_i \sin\zeta_j}{4\pi u_{ij}}\exp\left(ik\sqrt{1-u_{ij}^2}z_o\right)\exp\left(-\frac{W^2k^2u_{ij}^2}{4}\right)\begin{bmatrix} \cos\zeta_j \\ \cos\xi_i \\ \sqrt{1-u_{ij}^2} \end{bmatrix}.(12)$$

Then let us consider the propagation of each ijth plane wave through a uniaxially anisotropic medium. As shown in Fig. 4, the crystal axes (a, b, c) in the principal coordinate system are assumed to be oriented with respect to the xyz coordinate system such that their relation is described as

$$\begin{bmatrix} \hat{\mathbf{x}} \\ \hat{\mathbf{y}} \\ \hat{\mathbf{z}} \end{bmatrix} = \mathbf{R}\begin{bmatrix} \hat{\mathbf{a}} \\ \hat{\mathbf{b}} \\ \hat{\mathbf{c}} \end{bmatrix} = \begin{bmatrix} \cos\phi_c\cos\theta_c & -\sin\phi_c & \cos\phi_c\sin\theta_c \\ \sin\phi_c\cos\theta_c & \cos\phi_c & \sin\phi_c\sin\theta_c \\ -\sin\theta_c & 0 & \cos\theta_c \end{bmatrix}\begin{bmatrix} \hat{\mathbf{a}} \\ \hat{\mathbf{b}} \\ \hat{\mathbf{c}} \end{bmatrix}, \qquad (13)$$

where θ_c and ϕ_c are the angle between the c axis and the z axis, and the angle between the projection of the c axis on the xy plane and the x axis, respectively.

For the incident ijth plane wave $\mathbf{E}_{ij}^{inc}(x,y,z)$ given by Eqs. (11) and (12), we can determine eigenmodes of propagation inside the uniaxial crystal from the well-known wave equation in the momentum space:

$$\mathbf{k}\times\mathbf{k}\times\mathbf{E} + \omega^2\mu\tilde{\varepsilon}\mathbf{E} = 0, \qquad (14)$$

where $\mathbf{k}$ is the wave vector, μ is the magnetic permeability, and $\tilde{\varepsilon}$ is the dielectric permittivity tensor of the uniaxial crystal. From the boundary conditions at the medium surfaces, it is clear that the x and y components of the wave vector inside the medium remain the same as those of the ijth incident wave (i.e., $k\cos\xi_i$ and $k\cos\zeta_j$) provided it is homogenous in the xy plane and the interfaces are also parallel to the xy plane. The z components of the wave vectors in the medium can be obtained from the normal surfaces for the ordinary and extraordinary waves that are, respectively, given in the principal coordinate by

$$\frac{k_o^2}{n_o^2} = \frac{\omega^2}{c^2},$$

$$\frac{\alpha_e^2 + \beta_e^2}{n_e^2} + \frac{\gamma_e^2}{n_o^2} = \frac{\omega^2}{c^2}, \tag{15}$$

where n_o and n_e are the refractive indices of the ordinary and extraordinary waves, respectively, $k_o^2 = |\mathbf{k}_o|^2 = |\alpha_o\hat{\mathbf{a}} + \beta_o\hat{\mathbf{b}} + \gamma_o\hat{\mathbf{c}}|^2$, and $k_e^2 = |\mathbf{k}_e|^2 = |\alpha_e\hat{\mathbf{a}} + \beta_e\hat{\mathbf{b}} + \gamma_e\hat{\mathbf{c}}|^2$. These wave-vector in the (a, b, c) coordinate system can be written as

$$\begin{bmatrix} \alpha_o \\ \beta_o \\ \gamma_o \end{bmatrix}_{abc} = \mathbf{R}^{-1} \begin{bmatrix} k\cos\xi_i \\ k\cos\zeta_j \\ k_{oz,ij} \end{bmatrix}_{xyz},$$

$$\begin{bmatrix} \alpha_e \\ \beta_e \\ \gamma_e \end{bmatrix}_{abc} = \mathbf{R}^{-1} \begin{bmatrix} k\cos\xi_i \\ k\cos\zeta_j \\ k_{ez,ij} \end{bmatrix}_{xyz}, \tag{16}$$

with

$$\mathbf{R}^{-1} = \begin{bmatrix} \cos\phi_c\cos\theta_c & \sin\phi_c\cos\theta_c & -\sin\theta_c \\ -\sin\phi_c & \cos\phi_c & 0 \\ \cos\phi_c\sin\theta_c & \sin\phi_c\sin\theta_c & \cos\theta_c \end{bmatrix} \tag{17}$$

Substituting Eqs. (16) into Eqs. (15) leads to [5]

$$k_{oz,ij} = k\sqrt{n_o^2 - u_{ij}^2},$$

$$k'_{oz,ij} = -k_{oz,ij},$$

$$k_{ez,ij} = \frac{v + \sqrt{v^2 - 4uw}}{2u},$$

$$k'_{ez,ij} = \frac{v - \sqrt{v^2 - 4uw}}{2u}, \tag{18}$$

with

$$u = \frac{\sin^2 \theta_c}{n_e^2} + \frac{\cos^2 \theta_c}{n_o^2},$$

$$v = ku_{ij} \sin(\phi_c + \delta_{ij}) \sin 2\theta_c \left(\frac{1}{n_e^2} - \frac{1}{n_o^2} \right),$$

$$w = k^2 \left[\frac{u_{ij}^2}{n_e^2} - u_{ij}^2 \left(\frac{1}{n_e^2} - \frac{1}{n_o^2} \right) \sin^2(\phi_c + \delta_{ij}) \sin^2 \theta_c - 1 \right], \tag{19}$$

where a new variable δ_{ij} is introduced for convenience as

$$\delta_{ij} = \tan^{-1} \left(\frac{\cos \xi_i}{\cos \zeta_j} \right). \tag{20}$$

Here $k_{oz,ij}$, $k'_{oz,ij}$ correspond to the ordinary waves traveling toward the $+z$ and $-z$ directions, and $k_{ez,ij}$, $k'_{ez,ij}$ correspond to the extraordinary waves traveling toward the $+z$ and $-z$ directions, respectively. Note that these four z components of wave vectors in the birefringent plate are all real regardless of the c axis orientation and the incident angle provided the incident plane wave is nonevanescent ($u_{ij}^2 < 1$), and both the ordinary and extraordinary refractive indices are real and larger than a unity (n_o, $n_e > 1$). Thus, the wave vectors of the ordinary and extraordinary waves in the uniaxial crystal for the ijth plane wave can be expressed in the xyz coordinate system as

$$\mathbf{k}_{o,ij} = k \cos \xi_i \hat{\mathbf{x}} + k \cos \zeta_j \hat{\mathbf{y}} + k_{oz,ij} \hat{\mathbf{z}},$$

$$\mathbf{k}'_{o,ij} = k \cos \xi_i \hat{\mathbf{x}} + k \cos \zeta_j \hat{\mathbf{y}} + k'_{oz,ij} \hat{\mathbf{z}},$$

$$\mathbf{k}_{e,ij} = k \cos \xi_i \hat{\mathbf{x}} + k \cos \zeta_j \hat{\mathbf{y}} + k_{ez,ij} \hat{\mathbf{z}},$$

$$\mathbf{k}'_{e,ij} = k \cos \xi_i \hat{\mathbf{x}} + k \cos \zeta_j \hat{\mathbf{y}} + k'_{ez,ij} \hat{\mathbf{z}}. \tag{21}$$

From Eq. (14), we can also obtain the polarization states of these waves that are described in the principal coordinate as

$$\mathbf{o} = \begin{bmatrix} -\beta_o \\ \alpha_o \\ 0 \end{bmatrix}_{abc},$$

$$\mathbf{e} = \begin{bmatrix} -\alpha_e \gamma_e \\ -\beta_e \gamma_e \\ k_o^2 - \gamma_e^2 \end{bmatrix}_{abc}. \tag{22}$$

Note that $\mathbf{o}$, $\mathbf{e}$ are not unit vectors. Substituting each wave vector component in Eqs. (16) for Eqs. (22) and using the coordinate transformation given by Eq. (13), the polarization vectors can be expressed, in the xyz coordinate system, as

$$\hat{\mathbf{o}}_{ij} = N_{o,ij} \begin{bmatrix} -k\cos\zeta_j \cos\theta_c + k_{oz,ij}\sin\phi_c\sin\theta_c \\ k\cos\xi_i \cos\theta_c - k_{oz,ij}\cos\phi_c\sin\theta_c \\ ku_{ij}\cos(\phi_c + \delta_{ij})\sin\theta_c \end{bmatrix}_{xyz},$$

$$\hat{\mathbf{o}}'_{ij} = N'_{o,ij} \begin{bmatrix} -k\cos\zeta_j \cos\theta_c + k'_{oz,ij}\sin\phi_c\sin\theta_c \\ k\cos\xi_i \cos\theta_c - k'_{oz,ij}\cos\phi_c\sin\theta_c \\ ku_{ij}\cos(\phi_c + \delta_{ij})\sin\theta_c \end{bmatrix}_{xyz},$$

$$\hat{\mathbf{e}}_{ij} = N_{e,ij} \begin{bmatrix} -k_{ez,ij}k\cos\xi_i \cos\theta_c - \{k^2 u_{ij}\cos\xi_i \sin(\phi_c + \delta_{ij}) - k_o^2\cos\phi_c\}\sin\theta_c \\ -k_{ez,ij}k\cos\zeta_j \cos\theta_c - \{k^2 u_{ij}\cos\zeta_j \sin(\phi_c + \delta_{ij}) - k_o^2\sin\phi_c\}\sin\theta_c \\ \{k_o^2 - (k_{ez,ij})^2\}\cos\theta_c - k_{ez,ij}ku_{ij}\sin(\phi_c + \delta_{ij})\sin\theta_c \end{bmatrix}_{xyz},$$

$$\hat{\mathbf{e}}'_{ij} = N'_{e,ij} \begin{bmatrix} -k'_{ez,ij}k\cos\xi_i \cos\theta_c - \{k^2 u_{ij}\cos\xi_i \sin(\phi_c + \delta_{ij}) - k_o^2\cos\phi_c\}\sin\theta_c \\ -k'_{ez,ij}k\cos\zeta_j \cos\theta_c - \{k^2 u_{ij}\cos\zeta_j \sin(\phi_c + \delta_{ij}) - k_o^2\sin\phi_c\}\sin\theta_c \\ \{k_o^2 - (k'_{ez,ij})^2\}\cos\theta_c - k'_{ez,ij}ku_{ij}\sin(\phi_c + \delta_{ij})\sin\theta_c \end{bmatrix}_{xyz}, \tag{23}$$

where $N_{o,ij}$, $N'_{o,ij}$, $N_{e,ij}$, and $N'_{e,ij}$ are the normalization constants such that $\hat{\mathbf{o}}_{ij} \cdot \hat{\mathbf{o}}_{ij} = \hat{\mathbf{o}}'_{ij} \cdot \hat{\mathbf{o}}'_{ij} = \hat{\mathbf{e}}_{ij} \cdot \hat{\mathbf{e}}_{ij} = \hat{\mathbf{e}}'_{ij} \cdot \hat{\mathbf{e}}'_{ij} = 1$.

With four eigenmodes inside the uniaxial crystal derived so far, let us look into the propagation characteristics of the ijth plane-wave component $\mathbf{E}_{ij}^{inc}(x, y, z)$ through the birefringent medium, which has a thickness of d and its front and rear surfaces are in the xy plane, using the extended Jones matrix method [4,5]. Consider that the ijth plane wave $\mathbf{E}_{ij}^{inc}(x,y,z)$ traveling toward the $+z$ direction approaches to the medium and generates a partially reflected wave and a partially transmitted wave at the input surface of the plate located at $z = 0$. Thus, the electric fields of the incident, reflected, and refracted waves near the input face can be written, respectively, as

$$
\begin{aligned}
\mathbf{E}_{ij}^{inc}(x,y,z) &= \left(A_{s,ij}\hat{\mathbf{s}}_{ij} + A_{p,ij}\hat{\mathbf{p}}_{ij}\right)\exp\left(-i\mathbf{k}_{ij}\cdot\mathbf{r}\right), \\
\mathbf{E}_{ij}^{B}(x,y,z) &= \left(B_{s,ij}\hat{\mathbf{s}}'_{ij} + B_{p,ij}\hat{\mathbf{p}}'_{ij}\right)\exp\left(-i\mathbf{k}'_{ij}\cdot\mathbf{r}\right), \\
\mathbf{E}_{ij}^{C}(x,y,z) &= C_{o,ij}\hat{\mathbf{o}}_{ij}\exp\left(-i\mathbf{k}_{o,ij}\cdot\mathbf{r}\right) + C_{e,ij}\hat{\mathbf{e}}_{ij}\exp\left(-i\mathbf{k}_{e,ij}\cdot\mathbf{r}\right),
\end{aligned}
\tag{24}
$$

where $A_{s,ij}$, $A_{p,ij}$ are the amplitudes of the incident s and p waves, and $B_{s,ij}$, $B_{p,ij}$ are those of the reflected s and p waves, and $C_{o,ij}$, $C_{e,ij}$ are those of the transmitted o and e waves, respectively, at the input face. $\mathbf{k}'_{ij}$, $\hat{\mathbf{s}}'_{ij}$, and $\hat{\mathbf{p}}'_{ij}$ in the ijth reflected wave in Eqs. (24) can be expressed as

$$
\begin{aligned}
\mathbf{k}'_{ij} &= k\hat{\mathbf{u}}'_{ij} = k\left(\cos\xi_i\hat{\mathbf{x}} + \cos\zeta_j\hat{\mathbf{y}} - \sqrt{1-u_{ij}^2}\hat{\mathbf{z}}\right), \\
\hat{\mathbf{s}}'_{ij} &= \hat{\mathbf{s}}_{ij}, \\
\hat{\mathbf{p}}'_{ij} &= -\hat{\mathbf{u}}'_{ij}\times\hat{\mathbf{s}}_{ij} = \frac{1}{u_{ij}}\left(\sqrt{1-u_{ij}^2}\cos\xi_i\hat{\mathbf{x}} + \sqrt{1-u_{ij}^2}\cos\zeta_j\hat{\mathbf{y}} + u_{ij}^2\hat{\mathbf{z}}\right).
\end{aligned}
\tag{25}
$$

Similarly, the electric fields of the incident, reflected, and refracted waves near the output surface of the plate at $z = d$ are, respectively,

$$
\begin{aligned}
\mathbf{E}_{ij}^{C}(x,y,z) &= C_{o,ij}\hat{\mathbf{o}}_{ij}\exp\left(-i\mathbf{k}_{o,ij}\cdot\mathbf{r}\right) + C_{e,ij}\hat{\mathbf{e}}_{ij}\exp\left(-i\mathbf{k}_{e,ij}\cdot\mathbf{r}\right), \\
\mathbf{E}_{ij}^{D}(x,y,z) &= D_{o,ij}\hat{\mathbf{o}}'_{ij}\exp\left(-i\mathbf{k}'_{o,ij}\cdot\mathbf{r}\right) + D_{e,ij}\hat{\mathbf{e}}'_{ij}\exp\left(-i\mathbf{k}'_{e,ij}\cdot\mathbf{r}\right), \\
\mathbf{E}_{ij}^{trs}(x,y,z) &= \left(E_{s,ij}\hat{\mathbf{s}}_{ij} + E_{p,ij}\hat{\mathbf{p}}_{ij}\right)\exp\left(-i\mathbf{k}_{ij}\cdot\mathbf{r}\right),
\end{aligned}
\tag{26}
$$

where $C_{o,ij}$, $C_{e,ij}$ can be also defined as the amplitudes of the incident o and e waves, and $D_{o,ij}$, $D_{e,ij}$ are those of the reflected o and e waves, and $E_{s,ij}$, $E_{p,ij}$ are those of the transmitted s and p waves, respectively, at the output face. If multiple reflections inside the medium are negligible, the ijth electric fields in regions I, II, and III (See Fig. 1) can be written, respectively, as

$$
\begin{aligned}
\mathbf{E}_{ij}^{I}(x,y,z) &= \mathbf{E}_{ij}^{inc}(x,y,z) + \mathbf{E}_{ij}^{B}(x,y,z), \\
\mathbf{E}_{ij}^{II}(x,y,z) &= \mathbf{E}_{ij}^{C}(x,y,z) + \mathbf{E}_{ij}^{D}(x,y,z),
\end{aligned}
$$

Optoelectronic Materials and Technology

$$E_{ij}^{III}(x,y,z) = E_{ij}^{trs}(x,y,z). \qquad (27)$$

The corresponding magnetic fields can be derived by Maxwell's equation:

$$\mathbf{H} = \frac{i}{\omega\mu}\nabla \times \mathbf{E} = \frac{1}{\omega\mu}\mathbf{k}\times\mathbf{E}. \qquad (28)$$

Applying the boundary conditions that the tangential components of the electric and magnetic fields should be continuous at $z = 0$ and $z = d$, we can solve for eight amplitude coefficients, given by $A_{s,ij}$ and $A_{p,ij}$ given by Eq. (12). Note that instead of $\hat{\mathbf{x}}$ and $\hat{\mathbf{y}}$, the tangential xy plane can be also represented in terms of two coplanar orthogonal unit vectors $\hat{\mathbf{s}}_{ij}$ and $\hat{\mathbf{g}}_{ij}$ whose $\mathbf{E}$ and $\mathbf{H}$ components are used for the continuity conditions. Then, the following linear relations can be obtained:

$$\begin{bmatrix} C_{o,ij} \\ C_{e,ij} \end{bmatrix} = \mathbf{D}_{ij}^{in}\begin{bmatrix} A_{s,ij} \\ A_{p,ij} \end{bmatrix} = \begin{bmatrix} t_{so,ij} & t_{po,ij} \\ t_{se,ij} & t_{pe,ij} \end{bmatrix}\begin{bmatrix} A_{s,ij} \\ A_{p,ij} \end{bmatrix},$$

$$\begin{bmatrix} E_{s,ij} \\ E_{p,ij} \end{bmatrix} = \mathbf{D}_{ij}^{out}\mathbf{P}_{ij}\begin{bmatrix} C_{o,ij} \\ C_{e,ij} \end{bmatrix} = \begin{bmatrix} t_{os,ij} & t_{es,ij} \\ t_{op,ij} & t_{ep,ij} \end{bmatrix}\begin{bmatrix} \exp(-ik_{oz,ij}d) & 0 \\ 0 & \exp(-ik_{ez,ij}d) \end{bmatrix}\begin{bmatrix} C_{o,ij} \\ C_{e,ij} \end{bmatrix}, \qquad (29)$$

where $\mathbf{P}_{ij}$ is called the propagation matrix, and $\mathbf{D}_{ij}^{in}$ and $\mathbf{D}_{ij}^{out}$ the input and output dynamical matrices, respectively [4]. Two sets of the transmission coefficients ($t_{so,ij}$, $t_{po,ij}$, $t_{se,ij}$, $t_{pe,ij}$ and $t_{os,ij}$, $t_{es,ij}$, $t_{op,ij}$, $t_{ep,ij}$) at input and output surfaces, respectively, are given by

$$\mathbf{D}_{ij}^{in} = \begin{bmatrix} t_{so,ij} & t_{po,ij} \\ t_{se,ij} & t_{pe,ij} \end{bmatrix} = 2k\sqrt{1-u_{ij}^2}\begin{bmatrix} A & B \\ C & D \end{bmatrix}^{-1},$$

$$\mathbf{D}_{ij}^{out} = \begin{bmatrix} t_{os,ij} & t_{es,ij} \\ t_{op,ij} & t_{ep,ij} \end{bmatrix}$$

$$= \frac{\exp(ik\sqrt{1-u_{ij}^2}d)}{2k\sqrt{1-u_{ij}^2}}\left(\begin{bmatrix} A & B \\ C & D \end{bmatrix} - \begin{bmatrix} A' & B' \\ C' & D' \end{bmatrix}\begin{bmatrix} E' & F' \\ G' & H' \end{bmatrix}^{-1}\begin{bmatrix} E & F \\ G & H \end{bmatrix}\right), \qquad (30)$$

where

$$A = \hat{\mathbf{o}}\cdot(\hat{\mathbf{g}}_{ij}\times\mathbf{k}_{ij}) + \hat{\mathbf{o}}_{ij}\cdot(\hat{\mathbf{g}}_{ij}\times\mathbf{k}_{o,ij}), \qquad A' = \hat{\mathbf{o}}'\cdot(\hat{\mathbf{g}}_{ij}\times\mathbf{k}_{ij}) + \hat{\mathbf{o}}_{ij}'\cdot(\hat{\mathbf{g}}_{ij}\times\mathbf{k}_{o,ij}'),$$

$$B = \hat{\mathbf{e}}_{ij} \cdot \left(\hat{\mathbf{g}}_{ij} \times \mathbf{k}_{ij}\right) + \hat{\mathbf{e}}_{ij} \cdot \left(\hat{\mathbf{g}}_{ij} \times \mathbf{k}_{e,ij}\right),$$

$$C = k\hat{\mathbf{o}}_{ij} \cdot \hat{\mathbf{g}}_{ij} - \frac{\left(\hat{\mathbf{g}}_{ij} \times \mathbf{k}_{ij}\right) \cdot \left(\mathbf{k}_{o,ij} \times \hat{\mathbf{o}}_{ij}\right)}{k},$$

$$D = k\hat{\mathbf{e}}_{ij} \cdot \hat{\mathbf{g}}_{ij} - \frac{\left(\hat{\mathbf{g}}_{ij} \times \mathbf{k}_{ij}\right) \cdot \left(\mathbf{k}_{e,ij} \times \hat{\mathbf{e}}_{ij}\right)}{k},$$

$$E = 2\hat{\mathbf{o}}_{ij} \cdot \left(\hat{\mathbf{g}}_{ij} \times \mathbf{k}_{ij}\right) - A,$$

$$F = 2\hat{\mathbf{e}}_{ij} \cdot \left(\hat{\mathbf{g}}_{ij} \times \mathbf{k}_{ij}\right) - B,$$

$$G = 2k\hat{\mathbf{o}}_{ij} \cdot \hat{\mathbf{g}}_{ij} - C,$$

$$H = 2k\hat{\mathbf{e}}_{ij} \cdot \hat{\mathbf{g}}_{ij} - D,$$

$$B' = \hat{\mathbf{e}}'_{ij} \cdot \left(\hat{\mathbf{g}}_{ij} \times \mathbf{k}_{ij}\right) + \hat{\mathbf{e}}'_{ij} \cdot \left(\hat{\mathbf{g}}_{ij} \times \mathbf{k}'_{e,ij}\right),$$

$$C' = k\hat{\mathbf{o}}'_{ij} \cdot \hat{\mathbf{g}}_{ij} - \frac{\left(\hat{\mathbf{g}}_{ij} \times \mathbf{k}_{ij}\right) \cdot \left(\mathbf{k}'_{o,ij} \times \hat{\mathbf{o}}'_{ij}\right)}{k},$$

$$D' = k\hat{\mathbf{e}}'_{ij} \cdot \hat{\mathbf{g}}_{ij} - \frac{\left(\hat{\mathbf{g}}_{ij} \times \mathbf{k}_{ij}\right) \cdot \left(\mathbf{k}'_{e,ij} \times \hat{\mathbf{e}}'_{ij}\right)}{k},$$

$$E' = 2\hat{\mathbf{o}}'_{ij} \cdot \left(\hat{\mathbf{g}}_{ij} \times \mathbf{k}_{ij}\right) - A',$$

$$F' = 2\hat{\mathbf{e}}'_{ij} \cdot \left(\hat{\mathbf{g}}_{ij} \times \mathbf{k}_{ij}\right) - B',$$

$$G' = 2k\hat{\mathbf{o}}'_{ij} \cdot \hat{\mathbf{g}}_{ij} - C',$$

$$H' = 2k\hat{\mathbf{e}}'_{ij} \cdot \hat{\mathbf{g}}_{ij} - D'. \tag{31}$$

Therefore, $E_{s,ij}$ and $E_{p,ij}$ of the ijth transmitted plane wave $\mathbf{E}^{trs}_{ij}(x,y,z)$ are related with $A_{s,ij}$ and $A_{p,ij}$ of the ijth incident plane wave $\mathbf{E}^{inc}_{ij}(x,y,z)$ through a 2 × 2 transfer matrix $\mathbf{M}_{ij}$ as

$$\begin{bmatrix} E_{s,ij} \\ E_{p,ij} \end{bmatrix} = \mathbf{M}_{ij} \begin{bmatrix} A_{s,ij} \\ A_{p,ij} \end{bmatrix} = \mathbf{D}^{out}_{ij} \mathbf{P}_{ij} \mathbf{D}^{in}_{ij} \begin{bmatrix} A_{s,ij} \\ A_{p,ij} \end{bmatrix}. \tag{32}$$

It is interesting that the common phase factor $\exp\left(ik\sqrt{1-u_{ij}^2}\,d\right)$ in Eqs. (30) cannot be ignored contrary to the plane-wave case because it affects the transmission of the Gaussian beam similarly as $\exp\left(ik\sqrt{1-u_{ij}^2}\,z_o\right)$ term in Eq. (3). Under the assumptions that the system considered is linear and the principle of superposition of the electric fields is obeyed, we can obtain the total electric field of the transmitted Gaussian beam through the uniaxial medium from Eqs. (12) and (32):

$$\mathbf{E}^{trs}(x,y,z) = \sum_i \sum_j \mathbf{E}^{trs}_{ij}(x,y,z) = \sum_i \sum_j \left(E_{s,ij}\hat{\mathbf{s}}_{ij} + E_{p,ij}\hat{\mathbf{p}}_{ij}\right)\exp\left(-i\mathbf{k}_{ij} \cdot \mathbf{r}\right). \tag{33}$$

Furthermore, the above derivation is invariant of the position of the uniaxial plate, it can be simply applied to a series of N birefringent media by multiplying each $\mathbf{M}_{ij}$ given by Eq. (32) in sequence such that the total transfer matrix is $\mathbf{M}^T_{ij} = \mathbf{M}^N_{ij} \cdot \mathbf{M}^{N-1}_{ij} \cdots \mathbf{M}^2_{ij} \cdot \mathbf{M}^1_{ij}$, and summing up over i and j.

NUMERICAL CALCULATIONS

First, the longitudinal $|\mathbf{E}|^2$ distributions of the Gaussian beam propagating through an imaginary transparent isotropic medium having a refractive index of a unity in the xz plane ($y = 0$) are calculated as shown in Fig. 5. The total electric fields in regions I, II and III can be obtained from $\sum_i \sum_j \mathbf{E}_{ij}^I$, $\sum_i \sum_j \mathbf{E}_{ij}^{II}$, and $\sum_i \sum_j \mathbf{E}_{ij}^{III}$, respectively, whose ijth plane-wave components are given by Eqs. (27). The incident Gaussian beam is chosen to be normalized such that $|\mathbf{E}|^2 = 1$ at the beam waist ($E_o = 1$) and be focused at the input surface of the medium ($z_o = 0$). The other parameters are given in Table I. As expected, the Gaussian beam travels in the free space without any disturbance, and hence its natural characteristics such as the beam divergence can be observed. To model the fundamental-mode Gaussian beam accurately, the sufficient number of plane waves should be superimposed. In addition, the careful choice of ω_o and λ is necessary to guarantee that the contribution of the evanescent-wave components to the total electric field is negligible.

Next, look into the Gaussian beam propagating by way of an a-cut (or b-cut) uniaxial negative crystal where the c axis is perpendicular to the z axis (i.e., $\theta_c = 90°$) and $n_o > n_e$. Fig. 6 shows a typical example of the Gaussian beams partially propagating to the $+z$ and $-z$ direction throughout the birefringent plate. Here the c axis is supposed to be oriented to the $+x$ direction ($\phi_c = 0°$) and n_o and n_e are assumed to be 2 and 1.5, respectively. The $+z$ traveling beam [Fig. 6 (a)] corresponds to $\sum_i \sum_j \mathbf{E}_{ij}^{inc}$, $\sum_i \sum_j \mathbf{E}_{ij}^{C}$, and $\sum_i \sum_j \mathbf{E}_{ij}^{trs}$ in region I, II, and III, respectively, and the $-z$ traveling beam [Fig. 6 (b)] corresponds to $\sum_i \sum_j \mathbf{E}_{ij}^{B}$ and $\sum_i \sum_j \mathbf{E}_{ij}^{D}$, respectively. These two components comprise the entire Gaussian beam propagating through the medium and yield to the interference effects in regions I and II as shown in Fig. 7. It is noteworthy that these fine interference structures are similar with those of the plane-wave case as shown in Fig. 8. Although it is not clearly observable in Fig. 7 (a), the electric fields are discontinuous at the front boundary since the multiple reflections inside the medium are not included in the continuity equations. Strictly speaking, the electric fields do not need to be continuous at both the input and output faces of the plate because the boundary conditions do not consider the z components of the electric fields whose magnitudes are negligible compared with those of the total electric fields.

When the sample remains as a-cut (or b-cut), say, $\theta_c = 90°$, the cross-sectional $|\mathbf{E}|^2$ profile at the output face exhibits the apparent dependency on the c axis orientation as shown in Fig. 9 and the corresponding excited eigenmodes in the birefringent medium are calculated in Fig. 10. The incident circular Gaussian beam emerges as an elliptic Gaussian beam after interaction with the uniaxial crystal depending on the c axis orientation in the xy plane, and its major axis becomes perpendicular to the c axis. However, the extraordinary beam is almost extinct when the c axis is oriented perpendicular to the polarization of the incident Gaussian beam ($\phi_c = 90°$). Thus, the circular transverse pattern of the Gaussian beam is not altered through the medium as shown in Fig. 9 (b). As noticed from Fig. 10, only an extraordinary or ordinary mode is excited inside the medium when the crystal c axis is perpendicular or parallel to the polarization direction of the incident Gaussian beam, respectively. For any other crystal orientations in the xy plane, the internal Gaussian beam consists of both the o-beam and e-beam. In the case of a positive uniaxial medium, the elongate direction of the elliptic Gaussian beam becomes parallel to the c axis. The eccentricity of the emerging elliptical Gaussian beam increases with the sample thickness d_o and birefringence $|n_o - n_e|$, and decreases with the minimum spot size W_o. Note that n_o and n_e are chosen here to dramatize the effect, though most of practical uniaxial materials have small birefringence (i.e., $|n_o - n_e| \ll n_o, n_e$).

Fig. 11 illustrates evolution of the Gaussian beam through the uniaxial medium where the c axis is oriented along the <111> direction ($\theta_c = 54.736°$, $\phi_c = 45°$). In analogy with the plane-wave case, the incident Gaussian beam experiences double refraction and these refracted beams travel in different directions in the crystal [7]. Since the normal surface of the ordinary mode is a sphere, the o-beam propagates along the z axis and sustains the circular Gaussian beam. On the other hand, the normal surface of the extraordinary mode is nonspherical and is no longer normal to the plate surface for the oblique incidence. The e-beam deviates gradually from the z axis and is converted to the elliptic Gaussian beam. Once these two beams leave the plate, they both propagate normal to the crystal boundaries in the free space with a constant lateral distance of separation as shown in Fig. 11 (e) and (f).

In addition, the derived formulation can be applied to absorbing uniaxial media with the complex refractive indices n_o and n_e. Fig. 12 shows the transverse (at the $z = d$ plane) and longitudinal (at the $y = 0$ plane) $|E|^2$ distributions of the Gaussian beam propagating through a kind of o-type polarizer [4] with various c axis orientations in the xy plane ($\theta_c = 90°$). That is, the plate is assumed to have a real ordinary refractive index ($n_o = 1.5$) and a complex extraordinary refractive index ($n_e = 1.5 - i\ 1.5 \times 10^{-2}$) so that it transmits ordinary waves and attenuates extraordinary waves. Note that the reflected Gaussian beams at the medium surfaces are not calculated in Fig. 12 for simplicity. The transmitted Gaussian beam retains its circular transverse pattern regardless of the crystal orientations in the xy plan and it follows the general characteristics of the polarizer whose transmission axis is perpendicular to the c axis in the xy plane. In other words, the transmitted $|E|^2$ increases as ϕ_c changes from $0°$ to $90°$. When the c axis is perpendicular to the polarization of the incident Gaussian beam, the medium excites only the o-beam and it propagates through the crystal without additional losses except those resulting from Fresnel reflections as the case of the transparent medium (Compare with Fig. 13). When the c axis becomes parallel to the polarization of the incident Gaussian beam, the transmittance through the medium approaches zero because the excited e-beam is mostly absorbed in the medium.

CONCLUSIONS

Combining the angular spectrum of plane-wave technique with the extended Jones matrix method, the propagation of a linearly polarized Gaussian beam through a uniaxially anisotropic medium has been modeled and calculated. The numerical results obtained show various interesting features such as the conversion of the extraordinary ray from a circular to an elliptic Gaussian beam and double refraction depending on the crystal orientation. The reported model and the analytical approach are versatile and can be applied to a wide range of multistage birefringent plate, whether the medium is transparent or absorbing. With minor modifications, the model can be generalized to deal with the problem of any uniform anisotropic media including biaxial and gyrotropic materials interacting with a beam having an arbitrary waveform. Furthermore, most of the practical birefringent materials have small birefringence (i.e., $|n_o - n_e| \ll n_o, n_e$) and under this condition [4,5], the extended 2×2 Jones matrix formulation for Gaussian beam incidence can be greatly simplified to require much less computation time and to yield with almost identical calculation results.

REFERENCES

[1] R.C. Jones, "A New Calculus for the Treatment of Optical Systems. I. Description and Discussion of the Calculus," *Journal of Optical Society of America*, **31** 488–500 (1941).

[2] Pochi Yeh, "Electromagetic Propagation in Birefringent Media," *Journal of Optical Society of America*, **69** [5] 742–756 (1979).

[3] Pochi Yeh, "Optics of Anisotropic Layered Media: A New 4×4 Matrix Algebra," *Surface Science*, **96** 41–53 (1980).

[4] Pochi Yeh, "Extended Jones Matrix Method," *Journal of Optical Society of America*, **72** [4] 507–513 (1982).

[5] Claire Gu and Pochi Yeh, "Extended Jones Matrix Method. II," *Journal of Optical Society of America A*, **10** [5] 966–973 (1993).

[6] E.E.M. Khaled, S.C. Hill, and P. W. Barber, "Scattered and Internal Intensity of a Sphere Illuminated with a Gaussian Beam," *IEEE Transactions on Antennas and Propagation*, **41** [3] 295–303 (1993).

[7] B.E.A. Saleh and M.C. Teich, "Polarization and Crystal Optics": pp. 221–222 in *Fundamentals of Photonics*, New York: Wiley 1991.

Table I. Parameters used in numerical calculations

Incident Gaussian beam		Uniaxial crystal	
Peak amplitude E_o	normalized($\doteqdot 1$)	Ordinary index n_o *	2
Minimum spot size W_o	2 μm	Extraordinary index n_e *	1.5
Wavelength λ	0.633 μm	Thickness d	30 μm
Focal point z_0	0 μm	c-axis orientation θ_c, ϕ_c	variable
Interval width p	π/101		
Number of plane waves	5100		

*Fig. 5: $n_o = n_e = 1$
*Fig. 12: $n_o = 1.5$, $n_e = 1.5 - i\,1.5 \times 10^{-2}$
*Fig. 13: $n_o = n_e = 1.5$

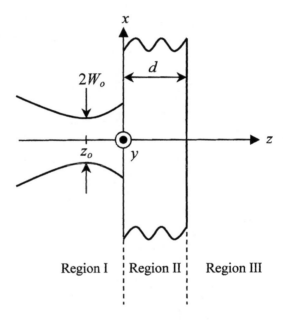

Fig. 1 Gaussian beam incident on a uniaxially anisotropic medium.

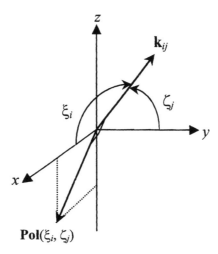

Fig. 2 Wave vector $\mathbf{k}_{ij}$ and polarization vector $\mathbf{Pol}(\xi_i, \zeta_j)$ of the ijth plane-wave component in the xyz coordinate system.

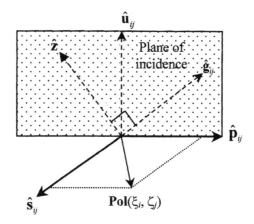

Fig. 3 Definitions of the plane of incidence, $\hat{\mathbf{s}}_{ij}$, and $\hat{\mathbf{p}}_{ij}$ for the ijth plane wave.

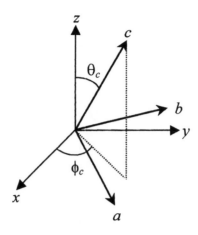

Fig. 4 Orientation of the crystal axes relative to the *xyz* coordinate system.

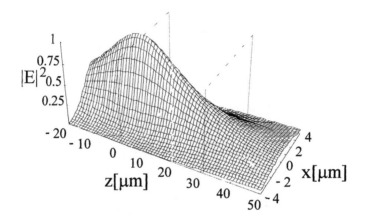

Fig. 5 Longitudinal $|\mathbf{E}|^2$ distributions of the Gaussian beam propagating through an imaginary isotropic medium with a refractive index of a unity in the *xz* plane.

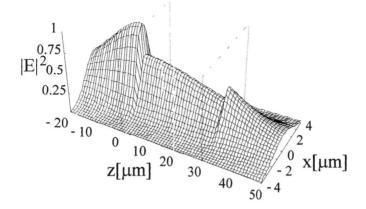

(a) Gaussian beam partially propagating to the + z direction.

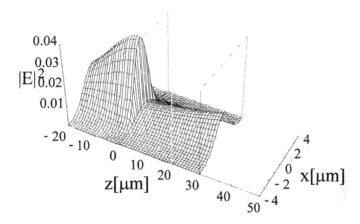

(b) Gaussian beam partially propagating to the − z direction.

Fig. 6 Longitudinal $|\mathbf{E}|^2$ distributions of the Gaussian beams partially propagating to the + z and − z direction throughout an a-cut (or b-cut) negative uniaxial medium ($\theta_c = 90°$, $\phi_c = 0°$) in the xz plane.

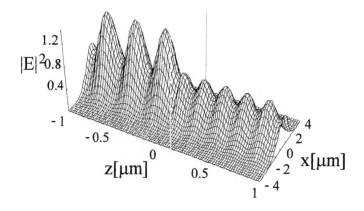

(a) $|\mathbf{E}|^2$ near the input crystal surface at $z = 0$ µm.

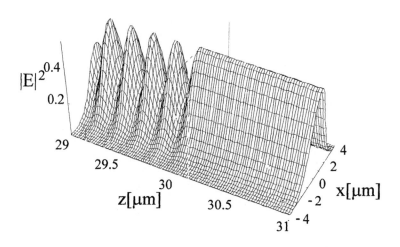

(b) $|\mathbf{E}|^2$ near the output crystal surface at $z = 30$ µm.

Fig. 7 Longitudinal $|\mathbf{E}|^2$ distributions of the entire Gaussian beam propagating through an a-cut (or b-cut) negative uniaxial medium ($\theta_c = 90°$, $\phi_c = 0°$) in the xz plane.

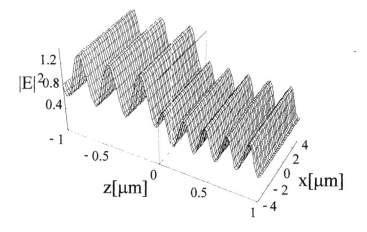

(a) $|\mathbf{E}|^2$ near the input crystal surface at $z = 0$ μm.

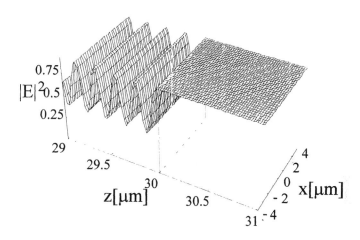

(b) $|\mathbf{E}|^2$ near the output crystal surface at $z = 30$ μm.

Fig. 8 Longitudinal $|\mathbf{E}|^2$ distributions of the plane wave propagating through an a-cut (or b-cut) negative uniaxial medium ($\theta_c = 90°$, $\phi_c = 0°$) in the xz plane.

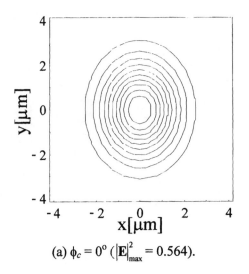

(a) $\phi_c = 0°$ ($|\mathbf{E}|^2_{max} = 0.564$).

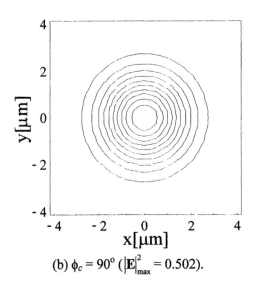

(b) $\phi_c = 90°$ ($|\mathbf{E}|^2_{max} = 0.502$).

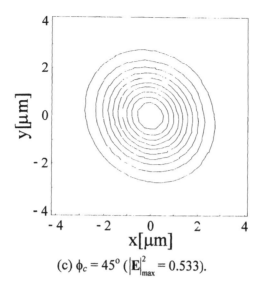

(c) $\phi_c = 45^\circ$ ($|\mathbf{E}|^2_{max} = 0.533$).

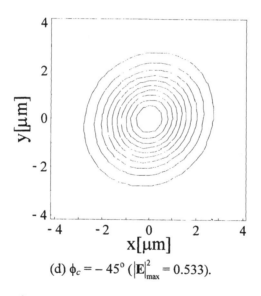

(d) $\phi_c = -45^\circ$ ($|\mathbf{E}|^2_{max} = 0.533$).

Fig. 9 Transverse $|\mathbf{E}|^2$ distributions of the Gaussian beam at the output surface of an a-cut (or b-cut) negative uniaxial medium ($\theta_c = 90^\circ$) in the xy plane. 10 equally spaced contour lines between 0 to $|\mathbf{E}|^2_{max}$ are drawn in each plot.

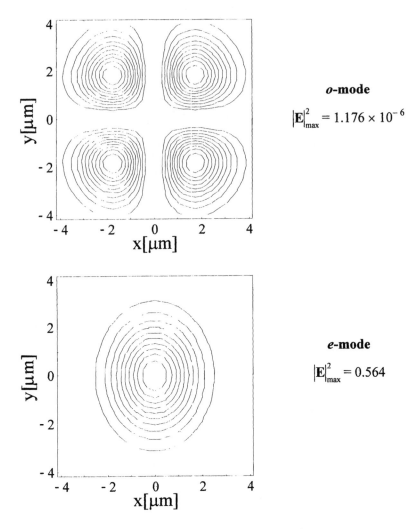

o-mode

$$|\mathbf{E}|^2_{max} = 1.176 \times 10^{-6}$$

e-mode

$$|\mathbf{E}|^2_{max} = 0.564$$

(a) Excited eigenmodes when $\phi_c = 0^o$.

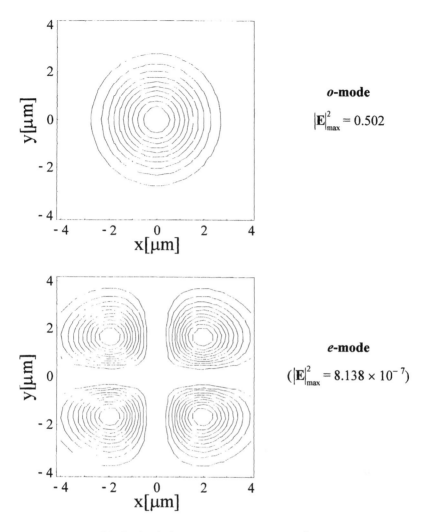

o-mode

$|\mathbf{E}|^2_{max} = 0.502$

e-mode

$(|\mathbf{E}|^2_{max} = 8.138 \times 10^{-7})$

(b) Excited eigenmodes when $\phi_c = 90°$.

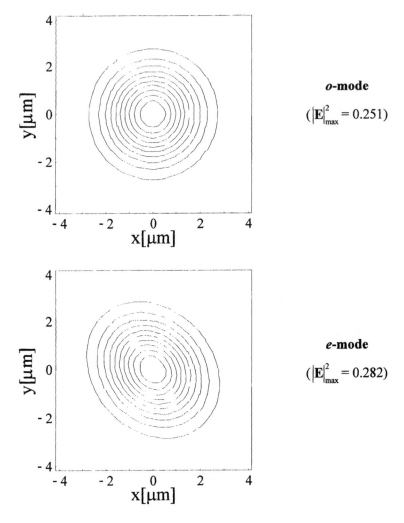

o-mode

$(|\mathbf{E}|_{max}^2 = 0.251)$

e-mode

$(|\mathbf{E}|_{max}^2 = 0.282)$

(c) Excited eigenmodes when $\phi_c = 45^o$.

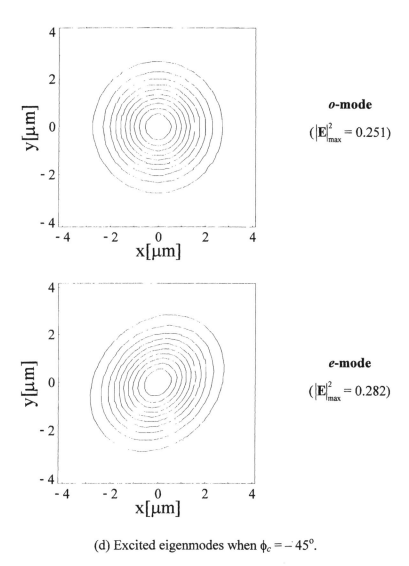

o-mode

$(|\mathbf{E}|^2_{max} = 0.251)$

e-mode

$(|\mathbf{E}|^2_{max} = 0.282)$

(d) Excited eigenmodes when $\phi_c = -45°$.

Fig. 10 Eigenmodes excitation in an a-cut (or b-cut) negative uniaxial medium (θ_c = 90°) at the output surface in the xy plane. 10 equally spaced contour lines between 0 to $|\mathbf{E}|^2_{max}$ are drawn in each plot.

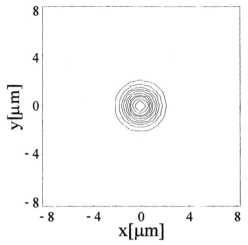

(a) $z = 0$ µm (at the input surface of the plate; $|\mathbf{E}|^2_{max} = 0.437$).

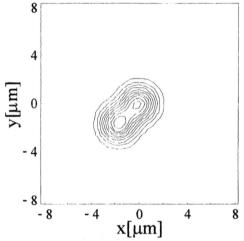

(b) $z = 10$ µm (inside the plate; $|\mathbf{E}|^2_{max} = 0.224$).

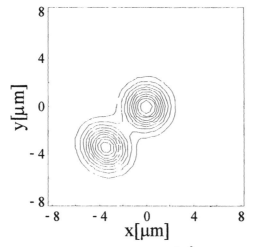

(c) z = 20 μm (inside the plate; $|\mathbf{E}|^2_{max} = 0.248$).

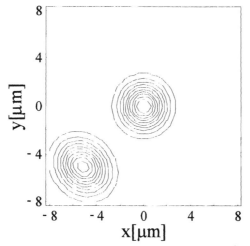

(d) z = 30 μm (at the output surface of the plate; $|\mathbf{E}|^2_{max} = 0.251$).

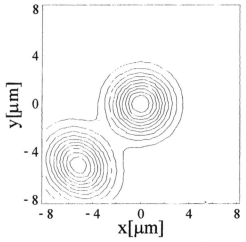

(e) $z = 40$ μm (after leaving the plate; $|\mathbf{E}|^2_{max} = 0.152$).

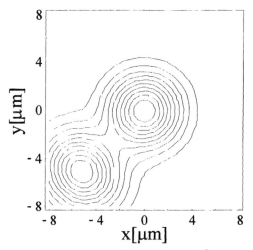

(f) $z = 50$ μm (after leaving the plate; $|\mathbf{E}|^2_{max} = 0.096$).

Fig. 11 Evolution of the Gaussian beam through a negative uniaxial medium where the c-axis is oriented the <111> direction ($\theta_c = 54.736°$, $\phi_c = 45°$) in the xy plane. 10 equally spaced contour lines between 0 to $|\mathbf{E}|^2_{max}$ are drawn in each plot.

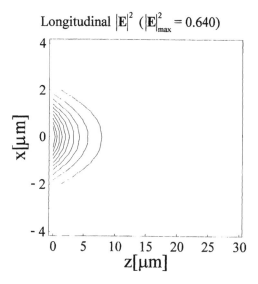

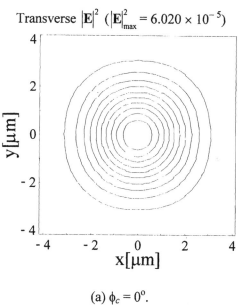

(a) $\phi_c = 0°$.

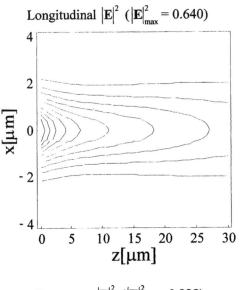

Longitudinal $|\mathbf{E}|^2$ ($|\mathbf{E}|^2_{max} = 0.640$)

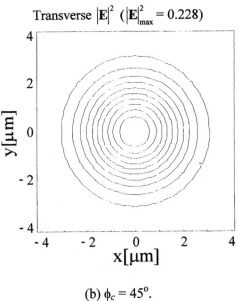

Transverse $|\mathbf{E}|^2$ ($|\mathbf{E}|^2_{max} = 0.228$)

(b) $\phi_c = 45°$.

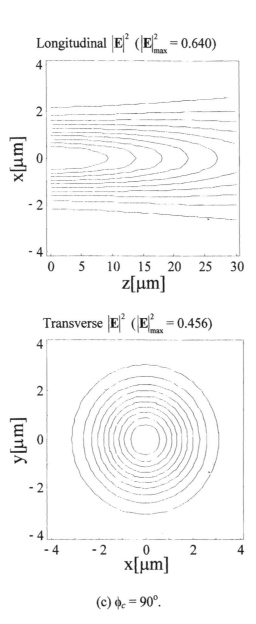

Longitudinal $|\mathbf{E}|^2$ ($|\mathbf{E}|^2_{max} = 0.640$)

Transverse $|\mathbf{E}|^2$ ($|\mathbf{E}|^2_{max} = 0.456$)

(c) $\phi_c = 90°$.

Fig. 12 Longitudinal (at the $y = 0$ plane) and transverse (at the $z = d$ plane) $|\mathbf{E}|^2$ distributions of the Gaussian beam propagating through an o-type polarizer ($n_o = 1.5$, $n_e = 1.5 - i\,1.5 \times 10^{-2}$).

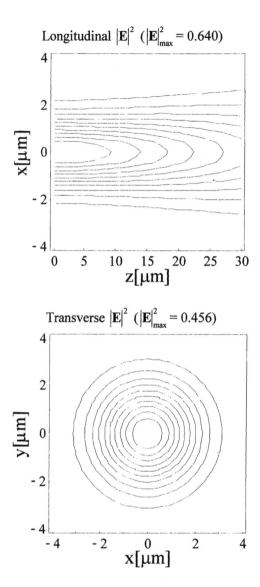

Fig. 13 Longitudinal (at the $y = 0$ plane) and transverse (at the $z = d$ plane) $|\mathbf{E}|^2$ distributions of the Gaussian beam propagating through a transparent isotropic medium ($n = 1.5$); Compare with Fig. 12 (c).

KEYWORD AND AUTHOR INDEX

Printed and bound by CPI Group (UK) Ltd, Croydon, CR0 4YY